Think Blockchain

A student's guide to blockchain's evolution from
Bitcoin, Ethereum, Hyperledger to Web3

Jerry Cuomo

DEDICATIONS

To: Rose, Gennaro, and Christophe—my students have become my teachers.

To: Future survivors. Royalties from this book are donated to the American Cancer Society.

TABLE OF CONTENTS

FOREWORD – Blockchain Paving the Internet Superhighway Towards Web3 ix

PREFACE xvii

CHAPTER 1– Blockchain Ducks 1

CHAPTER 2 – Three Blockchain Stories 21

CHAPTER 3 – How Blockchain Works 39

CHAPTER 4 – Cryptocurrency, Bitcoin, and Mining 59

CHAPTER 5 – Tokens, Ethereum, and Smart Contracts 87

CHAPTER 6 – Enterprise, Hyperledger Fabric, Modularity ... 121

CHAPTER 7 – Artificial Intelligence, IoT and Quantum 151

CHAPTER 8 – Cybersecurity, Zero-Knowledge Proofs, Digital Identity 163

CHAPTER 9 – A Road to Web3 195

CHAPTER 10 – Key TakeAways 225

ADDENDUM 1 – My Testimonies To Congress 235

ADDENDUM 2 – I'm Satoshi? 259

ACKNOWLEDGEMENTS 273

REFERENCES 281

Foreword –
BLOCKCHAIN PAVING THE INTERNET SUPERHIGHWAY TOWARDS WEB3

Web3 now aims to replace trust and good intentions with a block-chain-based network, where transparency and irrevocability are built into the technology. While it's not 100% clear yet of exactly how it will transpire. *Think Blockchain* is required reading for anyone who is looking to participate in shaping our industry's road to Web3. The future of the internet is in our hands.

– Foreword by Irving Wladawsky-Berger, Research Affiliate – MIT

...

From the first time I met Jerry in the hallways of IBM Research in Yorktown Heights, NY, it was clear that we shared the same passion for the importance of the rapidly evolving Internet. I believe our first conversation had to do with TCP/IP and Linux and the fact that multi-media applications were poised to transform the user experience of how consumers interact with businesses. We were just getting going.

Fast-forward to earlier this year, when I received an email from Jerry telling me that he enjoyed reading the Blockchain/Web3 entry I had posted on my blog and asked if I would be interested in writing the foreword for his upcoming book. Needless to say, I was thrilled with the opportunity... and here we are.

Think Blockchain is an enormously thoughtful and eminently practical guide for those curious to learn and gain experience on blockchain basics and its applications. Jerry's unique "story telling" style of writing, along with the accompanying illustrations, makes it easy to be an A+ student in his virtual blockchain class.

Jerry and I share a belief that blockchain will play a major role in the evolution of the Internet, although it is not 100% clear yet exactly how it will transpire. It's clear to many of us that the journey has begun and Web3 ideals are beginning to materialize.

Blockchains first came to light in 2008 as the architecture underpinning Bitcoin, the best known and most widely held cryptocurrency. It's a truly brilliant architecture built on decades-old fundamental research in cryptography, distributed data, game theory, and other advanced technologies. The blockchain's original vision was limited to enabling Bitcoin users to transact directly with each other with no need for a financial institution or government agency to certify the validity of the transactions. But, like the Internet and other transformative technologies, blockchain has now transcended its original objectives.

Blockchains are a kind of Distributed Ledger Technology (DLT). Over the past decade, an increasing number of people, including Jerry and I, consider blockchains and DLTs as major next steps in the evolution of the Internet. In 2016, the World Economic Forum (WEF) named The Blockchain in its annual list of Top Ten Emerging Technologies citing its potential to fundamentally change the way markets and governments work. "Like the Internet, the blockchain is an open, global infrastructure upon which other technologies and applications can be built," said the WEF. "And like the Internet, it allows people to bypass traditional intermediaries in their dealings with each other, thereby lowering or even eliminating transaction costs."

The Internet is a general-purpose network platform that unlocked huge innovations and economic value by significantly lowering the cost of connections and supporting a large variety of applications. A major reason for its ability to support so many different kinds of applications is that the Internet's foundation, its TCP-IP layer, has stuck to its basic data-transport mission, i.e., just moving bits around. It has no idea what the bits mean or what they're trying to accomplish.

The design decisions that shaped the Internet back in the mid-late 1980s didn't optimize for security and privacy, or for the ability to authenticate transactions between two or more parties. That's all the responsibility of the applications running on top of the TCP/IP layer, and they each generally do it their own way, sometimes not at all. Not surprisingly, the lack of standards for security, privacy, and transactional integrity has been one of the biggest challenges facing the Internet in our fast-growing digital economy.

Over the years, blockchain has transcended its original objectives and has evolved in two major directions. One continues to focus on blockchain as the underlying platform for Bitcoin, but it's also become the platform for the large number of cryptocurrencies, digital tokens, and other crypto-assets that have since been created. These topics are thoroughly covered by Jerry in Chapters 4, 5, and 9 in this book. The other objective is focused on the use of blockchain as a trusted distributed database for private and public sector applications involving multiple institutions, such as supply chains, financial services, and healthcare. This topic is covered in Chapters 2 and 6 of *Think Blockchain*.

Jerry's book provides a balanced view of how the cryptocurrency camp is based on public permissionless blockchains, which anyone can join and which require some kind of proof-of-work or proof-of-stake systems. The multi-institution camp is based primarily on private permissioned blockchains, where participation is restricted to the institutions transacting with each other.

Given my professional interests, I'm particularly interested in the use of blockchains in business and public sector applications for two major reasons. One reason why I consider blockchain as a major next step in the continuing evolution of the Internet is that it will help us enhance the security of Internet transactions and data by developing a layer with the standard encrypted services for secure communication, storage, and data access. And, the other reason is the fact that in the long-term, blockchain technologies can significantly improve the efficiency, resilience, and management of complex global applications involving multiple institutions.

But, while I've been somewhat skeptical of the crypto camp, a number of intriguing crypto-related topics have recently captured my attention, including Non-Fungible Tokens (NFTs), Decentralized Finance (DeFi), and especially Web3.

Significant amounts of capital, talent, and energy are now going into Web3 start-ups. "Venture capital firms have put more than $27 billion into crypto-related projects in 2021 alone—more than the 10 previous years combined—and much of that capital has gone to Web3 projects. And the industry has become a magnet for tech talent, with many employees of big tech firms quitting cushy, stable jobs to go seek their fortunes in Web3."

A good way of understanding Web3 is by comparing it to Web1 and Web2. Web1—the original Internet and World Wide Web of the 1990s and early 2000s—was primarily focused on publishing and accessing information on web pages using open protocols like HTTP. Web2 aka Web 2.0 emerged in 2005, as the next phase of the Internet, giving users the ability to create and publish their own content on personal websites, blogs, and social media platforms like Facebook, Twitter, and YouTube. Over time, most of this activity became dominated and monetized by a small number of global superstar companies.

There are multiple views on what Web3 is all about. Some critics view Web3 as little more than hype, a rebranding effort to shed some of the cultural and political baggage of crypto. Others believe it's a dystopian vision of a pay-to-play internet, in which every activity and social interaction becomes a financial instrument to be bought and sold.

On the other hand, its proponents argue that Web3 will replace today's corporate mega-platforms with blockchain-based networks that combine the open infrastructure of Web1 with the public participation of Web2, and that it will usher a more open, entrepreneurial Internet and a middleman-free digital economy. Supporters believe that Web3 will give creators and users a way to monetize their activity and contributions; that it will involve them in the governance and decision-making of the platforms supporting their work; and that it will give individuals more privacy and control over their data by being less reliant on advertising-based business models and targeted advertisements.

The original Internet aimed to develop a global decentralized computer network "in which no one need be in charge as long as everyone did their best to follow the same protocols and was tolerant of deviations. This system rapidly outcompeted all proprietary networks and changed the world. Unfortunately, time proved that the creators of this system were too idealistic, failing to take into account bad actors and, perhaps more importantly, failing to anticipate the enormous centralization of power that would be made possible by big data, even on top of a decentralized network," says Tim O'Reilly, founder of O'Reilly Media in a recent article titled, 'Why it's too early to get excited about Web3.'

Web3 now aims to replace trust and good intentions with a blockchain-based network, where transparency and irrevocability are built into the technology. "I love the idealism of the Web3 vision, but we've been there before," said O'Reilly. "During my career, we have gone through several cycles of decentralization and recentralization. The personal computer decentralized computing by providing a commodity PC architecture that anyone could build and that no one controlled. But Microsoft figured out how to recentralize the industry around a proprietary operating system. Open source software, the Internet, and the World Wide Web broke the stranglehold of proprietary software with free software and open protocols, but within a few decades, Google, Amazon, and others had built huge new monopolies founded on big data."

"Blockchain developers believe that this time they've found a structural answer to recentralization, but I tend to doubt it. An interesting question to ask is what the next locus for centralization and control might be. The rapid consolidation of Bitcoin mining into a small number of hands by way of lower energy costs for computation indicates one kind of recentralization. There will be others."

"If Web3 is to become a general-purpose financial system, or a general system for decentralized trust, it needs to develop robust interfaces with the real world, its legal systems, and the operating economy," adds O'Reilly. "The easy money to be made speculating on crypto assets seems to have distracted developers and investors from the hard work of building useful real-world services."

In conclusion, if Web3 heralds the birth of a new economic system, let's make it one that increases true wealth—not just paper wealth for those lucky enough to get in early but actual life-changing goods and services that make life better for everyone. The three blockchain stories that Jerry shares in Chapter 2 (Food safety, Identity theft prevention, and Anti-counterfeiting) are exactly the types of applications that illustrate how the additional trust brought by blockchain can improve our digital lives. However, we are not there yet. And the road to our future Internet will almost certainly be paved by those who are students of the state-of-the-art in computing, which is anchored by blockchain. The pages to follow in this book provide a sound blueprint for quickly coming up to speed on blockchain basics, from Bitcoin mining to Ethereum smart contracts, to Hyperledger with its modular, enterprise-ready design. When done you can add to your LinkedIn profile that you have experience writing a smart contract, building a digital token, and minting an NFT. *Think Blockchain* is required reading for anyone who is looking to participate in responsibly shaping our industry's road to Web3. The future of the Internet is in our hands. Let's shape it together.

PREFACE

Business or technical, the topics covered in this book will prove valuable to just about anyone curious and ready to become a scholar of blockchain.

WHO IS THIS BOOK FOR?

"A Student's Guide to Blockchain" was my original working title for this book. However, while testing this title with a few friends, we concluded that the topics covered in this book would be valuable to just about *anyone curious about the evolution of blockchain technology*. So, we replaced the word "student" with "thinkers" to welcome a broader audience. Now, "The Thinker's Guide to Blockchain" was somewhat of a mouthful, and after a brief conversation with my wife, Steph, we decided that "Think Blockchain" was it.

Also, THINK has been IBM's slogan for multiple generations – so what's not to like about thinking and/or being thoughtful about blockchain.

Think Blockchain is a practical guide to blockchain for students of all kinds. Blockchain skills are in high demand with blockchain job salaries booming over the past few years. Investopedia captures this thought well in an article titled "Forget Bitcoin: Blockchain is the Future," stating:

While Bitcoin and other cryptocurrencies grew intensely popular among the general financial and investment worlds in late 2017 and early 2018, they have since become more of a niche area for cryptocurrency enthusiasts. However, blockchain technology remains a quickly growing area of growth for compa-

nies across a host of industries. It is possible that blockchain technology will ultimately be seen as the most important innovation to come out of the cryptocurrency boom.[1]

The primary intent of *Think Blockchain* is to provide an entertaining and balanced education of the evolution of blockchain technology from Bitcoin to Ethereum, Hyperledger, Digital Assets, and Non-Fungible Tokens. The chapters are filled with unique images, many provided by designer Shaun Lynch, which complement the concepts in a unique way. The book combines both an aspirational and pragmatic overview of blockchain technology, its core capabilities, and the value they generate from a technical and business perspective. It describes various real-world examples, implementations, and approaches, with industry-specific and cross-industry use cases.

Developers will benefit from this book as several of the chapters are technical in nature, some even including simple coding exercises to further crystalize topics like blocks, chains, hashing, mining, smart contracts, and tokens.

Business leaders will equally benefit and are provided with a basis for understanding and evaluating how blockchain technology can transform their organization and business processes. I encourage business leaders to look over the simple coding examples. They won't bite and might better illustrate what makes blockchain tick.

Business or technical, the topics covered in this book will prove valuable to just about anyone curious and ready to become a scholar of blockchain.

WHAT IS COVERED IN THIS BOOK?

My goal is to cover all the key topics, which you need to gain a complete understanding of blockchain and how it has fueled the pioneers including Bitcoin, Ethereum, and Hyperledger. Even if you are already familiar with the basics, the early chapters will reinforce your understanding of the most important concepts and explore general use cases. As you dive deeper, you will be systematically introduced to specific topics with details and hands-on exercises that will enable you to successfully implement solutions that leverage blockchain.

Feel free, however, to jump directly to the chapter that most directly impacts your current role and answers your most immediate questions. You will also find references for further study throughout the chapters to fill in any gaps or provide more detail, depending on your level of experience or organizational role.

What follows is a short summary of each chapter in the book.

Chapter 1 – Blockchain Ducks

"If databases were birds, then blockchain would be a duck" is the thought that I share to kick off Chapter One, starting with this unique analogy for defining blockchain. The goal of this chapter is to give a broad definition of blockchain and set the tone, carried throughout this book, which is that while Bitcoin defined blockchain by putting cryptocurrency on the map, blockchain is an evolving technology poised to support thousands of use cases across all industries. Blockchain is defined first with is most universal features (called Blockchain DNA). It goes on to explain how permissionless and permissioned blockchains are the two dominant off-shoots bringing unique characteristics that are further described in the chapters to come.

Chapter 2 – Three Blockchain Stories

Blockchain is changing everyday life for the better for people like you and me. In Chapter Two, three of my favorite blockchain use cases are outlined, with Shaun Lynch's illustrations making the chapter perhaps the easiest read out of all the chapters in this book. The chapter covers how blockchain is being used for supply chain visibility by the Food Trust consortium to keep food safe—eliminating foodborne illnesses, reducing waste, and reducing dispute settlement times. The second story is focused on digital identity with the Verified.Me consortium establishing a foundation to reduce identity theft. And last is a story of counterfeit prevention and how cryptographic fingerprints of physical goods can be registered on blockchain as a proof of originality.

Chapter 3 – How Blockchain Works

Are you ready to create your own blockchain? Well, in this chapter, we're going to do just that. Before we get to the book's first coding exercise, I run through a brief history and evolution from currency—money—to today's blockchain-based currency, digital assets, and NFTs. I also provide a detailed description and examples of the fundamental technology elements of a blockchain, including blocks, chains, consensus, and distributed ledgers. Yes, and then we code, what might be, the world's smallest blockchain in under one-hundred lines of JavaScript code. If you're a coder, this will be a walk in the park, if not, please try to follow the example of "the Duckchain," it's easier than you might imagine, and it goes far to illustrate more vividly how blockchain works.

Chapter 4 - Cryptocurrency, Bitcoin, and Mining

A purely peer-to-peer version of electronic cash would allow online payments to be sent directly from one party to another without the burdens of going through a financial institution – Satoshi Nakamoto

This is the first chapter to review one of three blockchain pioneers. It's also the first of our triple-topic approach, where we take three intimately related topics. In this chapter's case, it's Cryptocurrency (use case), Bitcoin (technology), and Mining (breakthrough concept) and we show how each unique dimension plays into what makes this a pioneer. We discuss the evolution from fiat currency to crypto as an important primer on how the Bitcoin application creates an economy around a digital currency. I then outline the anatomy of a Bitcoin block and the process of mining it. I could not resist a coding example illustrating the mining process as a way to further exemplify pioneering aspects of Bitcoin.

Chapter 5 – Tokens, Ethereum, and Smart Contracts

Reports project that the market cap of the Non-Fungible-Token (NFT) industry will approach US$80 billion by 2025. Ethereum's smart contracts have ushered in the new, cool kids on the block in crypto. This chapter tackles the triple-topic of Tokens (use case), Ethereum (technology) and Smart Contracts (breakthrough concept). We start with another history lesson. This time, it's about my experience growing up (in NYC) using different types of tokens. From this, we create a taxonomy of digital tokens. Before creating a token or two, we deep-dive into Ethereum as a world super-computer and examine how smart contracts are the conceptual breakthrough that have enabled the token economy and allow the expansion of the scope of applications supported by blockchain. With this, another coding venture is attempted, in this case, creating an Ethereum smart contract and an ERC-20 token. We end the chapter with an exercise in creating a NFT token and a "buyer beware warning."

Chapter 6 – Enterprise, Hyperledger Fabric, Modularity

Fabric is like a Lego kit that allows you to assemble the pieces to produce the Millennium Falcon from *Star Wars*, or a model of the Eiffel Tower. Fabric has the potential energy to become just about any type of blockchain network, with "some assembly required." This chapter starts off with my recollection of the events that transpired that led to IBM's involvement in co-creating the Hyperledger project under the Linux Foundation. The premise of this chapter is that Hyperledger Fabric is a pioneer of blockchain because of how its modular architecture enables an extreme variety of blockchain use cases that are aligned with enterprise requirements. With that, the chapter takes a specific look at accountability (identity), privacy (channels), expressiveness of business logic (chain-code), and security and performance that comes from unique consensus models. The chapter concludes with a rich list of examples of enterprise blockchains beyond cryptocurrency as a means to spark your imagination to help you think about what you might create.

Chapter 7 – Artificial Intelligence, IoT and Quantum

The thought of bringing these ingredients together may be viewed by some as brewing a modern-day version of IT pixie dust. Let's let the magic begin. In this chapter, you learn how blockchain is at the nexus of emerging technology, which includes artificial intelligence, the Internet of Things, and quantum computing, and see how blockchain technology can add the missing element of trust into the picture. This includes using blockchain as an audit trail for training data of machine learning models and IoT firmware versions allowing users to quickly identify whether associated data is authentic and has not been tampered with. We also discuss how the highly touted security of blockchain is beginning to look like it is set to change with the emergence of powerful quantum computers. It will be child's play for such devices to break the kinds of cryptographic protection implemented in existing blockchain frameworks. But don't fret, there are a few good answers proposed in this chapter.

Chapter 8 - Cybersecurity, Zero Knowledge Proofs, Digital Identity

Blockchains possess ingredients that make it more naturally resistant to cyber attacks. The blockchain's inherently decentralized nature makes it a natural deterrent and well-suited for cybersecurity use cases. Whether building on its properties of advanced cryptography or decentralization, blockchain technology can thwart many of today's most determined cyber attacks. This chapter looks at a set of examples where blockchain can make a difference in improving cybersecurity. Topics covered include decentralized cloud services like Domain Name Services (DNS) and storage. Advancements in cryptography are highlighted through examples of the almost magic technology behind Zero Knowledge Proofs (ZKP) and the role a universal digital identity standard that gives individuals control of their digital identities. The chapter is rounded out with a look at how securing software supply chains with blockchain can avoid malware interrupting our digital economy.

Chapter 9 - A Road to Web3

Is Web3... the Internet of the future? A more decentralized internet? A "stateful internet" of value? Or is it just a buzzword? The Web3 topic can very well be the phenomenon that brings all the previous concepts covered in this book together in a more unified and sensible way. The general movement behind Web3 is the ideal of shifting more profit and control to data creators. The chapter illustrates that this sort of shift is not as radical as one might think and that we've seen this behavior before. With that, Web3 aims to wield blockchain's decentralization nature to temper the dominance of big tech over internet use and its control over personal data. In the chapter, Web3 is described as an extension to Web2, adding a new universal state layer where valuable, personal, and sensitive data is kept in a "neutral zone" outside the confidence of big organizations. While Web3 is still being formulated, its early applications are reviewed, including NFTs, DeFi, metaverse, blockchain oracles, and universal identity. The chapter concludes with a call to action by identifying and proposing solutions to existing issues that need to be addressed before Web3 can further mature and gain adoption over the decades to come.

Chapter 10 - Key Takeaways

This chapter includes a handful of takeaway points that are most worth remembering. Rather than summarize those points here, I thought these points were worth their own section in the Preface. So, hang on and they will be covered in the section that follows.

Addendum 1 – My Testimonies to Congress

This supplemental chapter contains three transcripts of my blockchain-related testimonies to the US Government. Two of the testimonies are to congressional committees hosted by the House Energy and Commerce Subcommittee. The other was a President Obama-initiated Commission on Enhancing National Cybersecurity. These transcripts were included in the book because I feel the thoughts contained within were a useful takeaway and could act as an informative summary for this book.

Addendum 2 - I'm Satoshi?

This supplemental chapter was written by my friend and colleague Mark Parzygnat. The idea for this chapter was tossed around during a conversation between Mark and me, as we reflected that not knowing who invented Bitcoin was a big deal, and unprecedented in modern times. When you think about it, we pretty much know who contributed to the introduction of all major advancements in technology. Mark reflects that perhaps not since the invention of fire or the wheel can we not track the ownership to a group or individual. In the chapter, a profile of Satoshi Nakamoto is established that looks and tries to match it against a few known candidates. Mark also boldly identifies who he thinks the real Satoshi is. Will you agree? The chapter takes a lighthearted approach to answering the question of who Satoshi is, while also looks at the important phenomenon created that launched decentralized cryptocurrency and got us all to "*Think Blockchain.*"

KEY TAKEAWAYS

Tell them what *you are going* to *tell them, tell them, then tell them* what *you told...*

I heard that this is the tried and true method for teaching and having the lessons stick. There are a few key thoughts repeated in this book. This section summarizes what likely are the most essential points to remember from *Think*

Blockchain. These points are also repeated in Chapter 10 with slightly more elaborate accompaniment.

If databases were birds, blockchain would be a duck

Blockchain's truly brilliant architecture is built on decades-old fundamental research in cryptography, distributed transaction processing, game theory, and yes—database technology too.

Bitcoin is just one of one thousand applications of blockchain

Applying blockchain technology broadly will transform industries—because its application can go far beyond cryptocurrency and Bitcoin. Blockchain technology comes in many shapes and sizes. Modern blockchain is flexible and ultra-efficient. Oh… and it doesn't always consume energy like Bitcoin does.

Decentralized blockchain establishes unprecedented trust

How much decentralizing is good enough to establish trust? Designing your blockchain application with a **Minimally Viable Decentralization (MVD)** is the key to balance performance and unprecedented trust.

Blockchain-based trust breeds wide-scale adoption

Blockchain's combination of reputational trust and algorithmic trust calms fear, trepidation and encourages wide-scale usage from a large, diverse population of users. When this trust model is applied correctly, a safe, secure, and vibrant ecosystem will quickly emerge.

Blockchain at the nexus of emerging technology

Trust is also essential for the wide-scale adoption of emerging technologies. With software and open source "eating the world," trusting technology is not an option; it is essential.

Web3 is the future internet fueled by blockchain

While many say Web3 is the future of the internet, skeptics say, it's just a buzzword. There is no doubt in my mind that the wonderful thing that we call internet will evolve. The next wave must tackle the toughest challenges covered in this book, however, doing so will provide a means to flow more control and rewards to those creating the value on the web.

Blockchain is changing everyday life for good

The final key takeaway is that blockchain is a technology that is poised to change everyday life for the better. It is here for good. The double meaning is that it is here to stay and to bring a new level of trust, built on reputational and cryptographic algorithms. A balanced view will almost certainly prevail. Hybrid blockchains are the responsible choice of network designers because they often possess a minimally viable decentralized design and are deployed to an interoperable web of public-permissioned networks. And yes, changing life for the better also means it's time the creators of data—whether that be an NFT, your healthcare record or your digital identity—be granted control and profit from that data. Therefore, the responsible development of blockchain is a team sport involving users, governments and businesses collaborating to both protect and liberate the world's data. With this, blockchain as a technology and a mindset breeds trustworthy computing and becomes the fuel that propels Web3 and the future of our internet – for good.

MEET THE AUTHOR

Photo – Sheldon the dog(left), Gwenevere the owl (left) and Jerry (right)

Gennaro (Jerry) Cuomo is an IBM Fellow and considered amongst the most prolific contributors to IBM's software business, producing products and technologies that have profoundly impacted how the industry conducts commerce over the World Wide Web. He is most recognized as one of the founding fathers of WebSphere Software, whose innovations defined WebSphere as the industry-leading application server serving over eighty thousand customers. At IBM, Cuomo has led projects in the areas of AI-powered automation, blockchain, APIs, cloud computing, mobile computing, Internet of Things, web server performance and availability, and service-oriented architecture.

Cuomo has filed for over a hundred US patents and has been cited over three thousand times. His most visible patent is the first use of the "Someone is typing…" indicator found in today's most popular instant messaging applications.

Recently, Cuomo and team have illustrated how blockchain can revolutionize business and redefine companies and economies with the creation of Linux Foundation's Hyperledger Fabric and the IBM Blockchain Platform. In March 2016 and February 2018, Cuomo was called upon by the United States government as an expert witness to testify to US Energy and Commerce Committee on Digital Currency and Blockchain. During his 2016 testimony, Cuomo urged the Obama administration to adopt blockchain as a primary means to protect citizen identity and to enhance national security.

Cuomo is the co-author of the book *Blockchain for Business*. He also recently authored the book *The Art of Automation,* and is the host of a podcast of the same name.

Cuomo enjoys playing golf and drinking craft beer (not necessarily at the same time) with his friends and best friend/wife Steph. He also enjoys walking his dogs and playing bass guitar in the band Mind the Gap.

UP NEXT | QUIZ TIME

The "Up Next" sections can be found at the conclusion of each chapter of this book and serve as a means to tie together the topics covered in various chapters. Also, given the educational nature of this book, I conclude many of the chapters with a set of questions to quiz you on the key takeaway themes of each chapter. Should be a piece of cake.

Oh, and for the curious coders out there, a few of the chapters have additional coding assignments that challenge you to extend the examples covered in the book in new and interesting ways. All coding examples covered in this book, along with the extended assignments, can be found in our Github repository called **Think Blockchain Labs.**

- - -

Now, fasten your seatbelts; the book is about to take off.

Chapter 1–
BLOCKCHAIN DUCKS

If databases were birds, then blockchain would be a duck.

COVERED IN THIS CHAPTER

- Blockchain defined

- Database analogy

- DNA of Blockchain

- Enterprise Blockchain defined

- Blockchain is not Bitcoin

HERE WE GO...

I thought it would be appropriate to start off with a description of blockchain. I have noticed such descriptions often wildly vary from person to person. I'm hoping mine will resonate and fuel the concepts shared in this book. As you will see in the chapters to follow, I like telling stories and drawing analogies. They are often fun to share and help the learning process. So, what better way to describe blockchain than with a story... or is it an analogy? I'm not sure but will let you decide.

You see, I am often asked to describe what blockchain is. After a few years of trying to perfect the description, I found what works best is if you start from a point that is well known and build your description from there, then you are less likely to lose your audience. (Wait, I haven't lost you yet, have I? It's only the second paragraph of this book!) I found comparing blockchain to **a database** provides that well-understood starting point by which a blockchain definition can be established. My saying would go something like:

If you think of blockchain as a database, you wouldn't be wrong. Now, you wouldn't be right either. Blah blah blah.

In front of a large audience in Cleveland Ohio,[2] I was planning to start off with such a short description of blockchain. Having done it before, seemingly hundreds of times, I decided to have a little fun with the description, and this is what I came up with. Later, when I shared the analogy with the gifted graphics designer, Shawn Lynch, he created these compelling visuals to go along.

IF DATABASES WERE BIRDS, THEN BLOCKCHAIN WOULD BE A DUCK

If databases were birds, you wouldn't be wrong to consider blockchain as a type of bird. Say a duck. That is, blockchains do have some database-like features. Just like ducks share many similar characteristics with other members of the bird family. (Btw... duck is the common name for numerous species of waterfowl in the family Anatidae. Just saying.) But they also have unique features not exhibited by other types of birds.

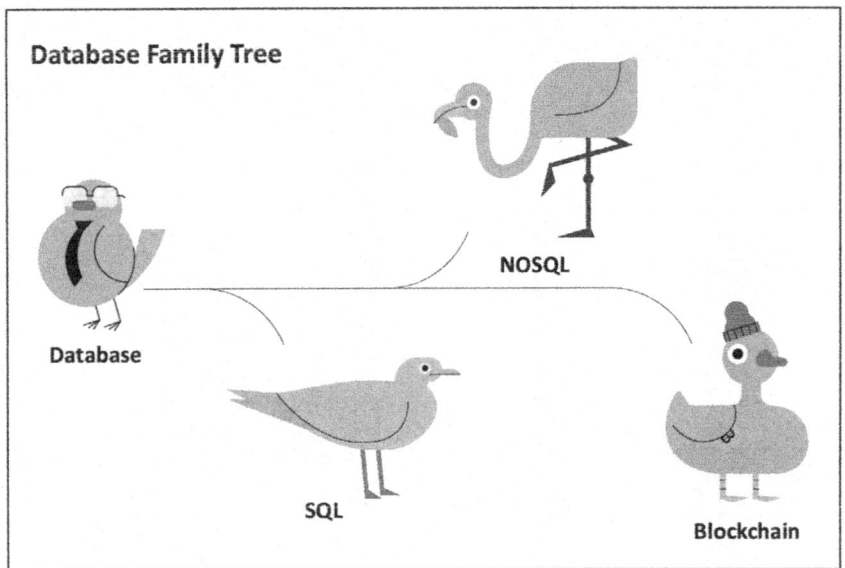

Chapter 1 – Figure 1 – Database-Bird Family Tree including Blockchain-Ducks

Starting with the common features, you might say there is a standard DNA shared by all database-ducks. You know, most birds have feathers, wings, bills, and many can fly. Similarly, most all database-birds have administrators, support data storage (i.e., a ledger) and reliable transaction processing (i.e., ACID – Atomicity, Consistency, Isolation, Durability properties[3]). A Database Management System (DBMS) functions using system commands, first receiving instructions from a database administrator in DBMS, then instructing the system accordingly, either to retrieve data, modify data, or load existing data from the system. Popular DBMS examples include cloud-based database management systems, In-Memory Database Management Systems (IMDBMS), Columnar Database Management Systems (CDBMS), and NoSQL in DBMS.

As proper members of the database-bird family, almost all blockchains support these very same concepts that you will see in the description to come. However, blockchain-ducks exhibit unique features from a standard database-bird. You know, most ducks waddle, have bills, and have webbed feet.

In a similar vein, all blockchain-ducks have **shared ledgers, consensus, and immutability**. The next sections explain these 3 concepts further.

Distributed Ledger

Chapter 1 – Figure 2 – Blockchain-Ducks distributed ledger across multiple administrators

Unlike database-birds, which have a single centralized administrator, who sets up the operational rules for the data ledger, a blockchain-duck has multiple administrators that each has an exact copy of the ledger.

A distributed ledger (also called a shared ledger or Distributed Ledger Technology or DLT) is a consensus of replicated, shared, and synchronized digital data geographically spread across multiple sites, countries, or institutions. Unlike with a centralized database, there is no central administrator.[4]

Consensus

Chapter 1 – Figure 3 – Blockchain-Ducks consent to validity of transactions before committing

Unlike a database-bird that when a transaction gets submitted it is immediately committed, provided it conforms to the operational rules set by the centralized database administrator, transactions in a blockchain network are first proposed to a group of (decentralized) administrators. Each administrator tests the fitness of the transaction against their copy of the shared ledger and shares their results with the group. If consensus is reached by the group that the transaction is fit, then and only then is the transaction committed and replicated to all the shared ledgers. Now, that's an ultra-simplified view of consensus that is sufficient for a basic definition of blockchain. In an upcoming chapter, we will cover consensus in more detail.

Immutability

Chapter 1 – Figure 4 – Blockchain-Ducks can't change a transaction, once committed

The last distinguishing characteristic of a blockchain-duck over a generic database-bird is immutability. While the database-bird administrator has access to the full gamut of commands including commands like update and delete that can change records in the ledger, blockchain-ducks cannot wield such power. Once a transaction is committed to a blockchain,

the blockchain-duck administrators don't have access to such commands. Blockchains are append-only. When a block is committed, it is cryptographically secured with other blocks in the ledger. This act transforms the shared ledger into an audit log of sorts that becomes the foundation of trust.

As I'm writing this down now, I'm realizing maybe I should have added a fourth characteristic shared by all blockchain-ducks—**cryptography**.

Cryptography

Chapter 1 – Figure 5 – Blockchain-Ducks are great at mathematics and cyphers

The term cryptography is derived from the Greek word *'kryptos'*, which means 'hidden'. A blockchain-duck is expert at using mathematical techniques to hide or scrabble data. As we will see in chapters to follow when we cover topics like hashing and mining, the integrity of a blockchain is gained through strong cryptography. Blockchains spend so much time performing the math to encrypt data that some view them as giant cryptographic engines. In fact, there is an entire industry dedicated to building specialized computer hardware processors that accelerate blockchain cryptography functions.

We will go over the mechanics of how this comes together in blocks and chains in an upcoming chapter. I'll continue to keep the concepts high-level to capture the key points.

So, all blockchain-ducks have these features in common. But as we know, not all ducks are exactly alike. The next section singles out a few of the popular styles of blockchains.

NOT ALL BLOCKCHAIN-DUCKS ARE THE SAME

There are many types of duck species, with apparently two primary types including dabbling ducks and diving ducks. Similarly, there are many species of blockchain—Bitcoin, Ethereum, R3-Corda, Ripple, Hyperledger Fabric— just to name a few.

Chapter 1 – Figure 6 – A sampling of Blockchain-Ducks species

In this section (like the two dominant duck types), we will focus on two dominant styles of blockchain—**permissionless** and **permissioned.**

PERMISSIONLESS OR PUBLIC BLOCKCHAINS

Permissionless or public blockchains possess the basic "DNA" of a blockchain described above, which means they are shared, cryptographically secured, and anchored by an immutable digital ledger. They also exhibit unique characteristics that align them well for certain types of applications, especially emulating cash via cryptocurrency. The sections that follow highlight several of these unique attributes for each of these styles, starting with the permissionless (or public) blockchain.

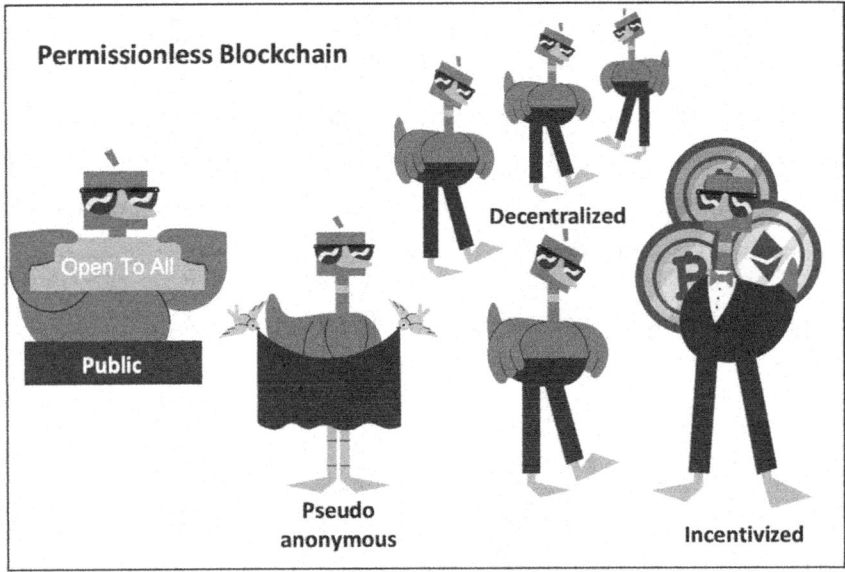

Chapter 1 – Figure 7 – Permissionless or Public Blockchains
are known to exhibit these characteristics

Public

Blockchains like Bitcoin and Ethereum are permissionless in nature, which means they allow anyone to join. Typically, public blockchains allow all nodes of the blockchain to have equal rights to access the blockchain, create new blocks of data, and validate blocks of data.

For example, the Bitcoin blockchain is designed as a decentralized peer-to-peer (P2P) network, where nobody owns or controls Bitcoin and everyone can take part. Bitcoin software is free to download and run. A Bitcoin node is a program that fully validates transactions and blocks. By running a Bitcoin node, you are supporting the Bitcoin network to become more decentralized and are fostering the growth of Bitcoin.

Pseudo anonymous

In Bitcoin, the identity of members engaging the network is noted under an alias. More accurately, members have a pseudonym, which is the address at which you receive bitcoin. Every transaction involving that address is stored forever in the blockchain. However, if I linked my digital address (pseudonym) to my real identity (Jerry Cuomo), then every transaction will be linked to me. Sending and receiving bitcoins is like writing a book under a pseudonym. Hence, *Think Blockchain* would be credited to my digital address and would look something like this:

Think Blockchain, by *3FZbgi29cpjq2GjdwV8eyHuJJnkLtktZc5*.

Decentralized

In blockchain, decentralization refers to the transfer of control and decision-making from a centralized entity (individual, organization, or group thereof) to a distributed network. Decentralization of data via a shared ledger is one of the keys to its trustworthiness. For example, decentralization makes the Bitcoin network secure and difficult for malicious people and hackers to penetrate or interfere with because they can't control the entire system.

Incentivized

Public blockchains used for exchanging and mining cryptocurrency take time and energy, which translates to money. Specifically, on these public blockchains, the nodes "mine" for cryptocurrency by creating blocks for the transactions requested on the network by solving cryptographic equations. In return for this hard work, the miner nodes earn a small amount of cryptocurrency. The miners essentially act as new stockbrokers that formulate a transaction and receive (or "mine") a commission (or transaction fee) for their efforts.

- - -

Permissionless networks were not initially known for their speed of processing transactions. But with each new generation, there are improvements. For example, Bitcoin processes 4.6 transactions per second. In comparison, Visa does around 1,700 transactions per second on average (based on a calculation derived from the official claim of over 150 million transactions per day).[5] However, like other permissionless and mining-based blockchains, Bitcoin is "slow" for a good reason. Its speed is related to the fact that records (known as blocks) in the Bitcoin blockchain are limited in size and frequency. The on-chain transaction processing capacity of the Bitcoin network is throttled (limited) by the average block creation time of 10 minutes and the original block size limit of 1 megabyte. In simple terms, the reason for this is to ensure fairness and to prevent "the administrator with the fastest computer from always winning."

Now that we've covered the key attributes of the "dabbling duck," let's take a closer look at the other type of blockchain-duck, which I guess we will have to call the "diving duck." (Are you bored of the duck analogy yet? Well, as you will see, we are just really getting started.)

PERMISSIONED OR ENTERPRISE BLOCKCHAIN

An enterprise blockchain—like Hyperledger Fabric and R3's Corda—exhibits four key enhancements to blockchain's base DNA as shown in the figure that follows. Permissioned blockchain can also inherit features of permissionless blockchains—they support aspects of decentralization, can be made public, and employ unique strategies to incentivize.

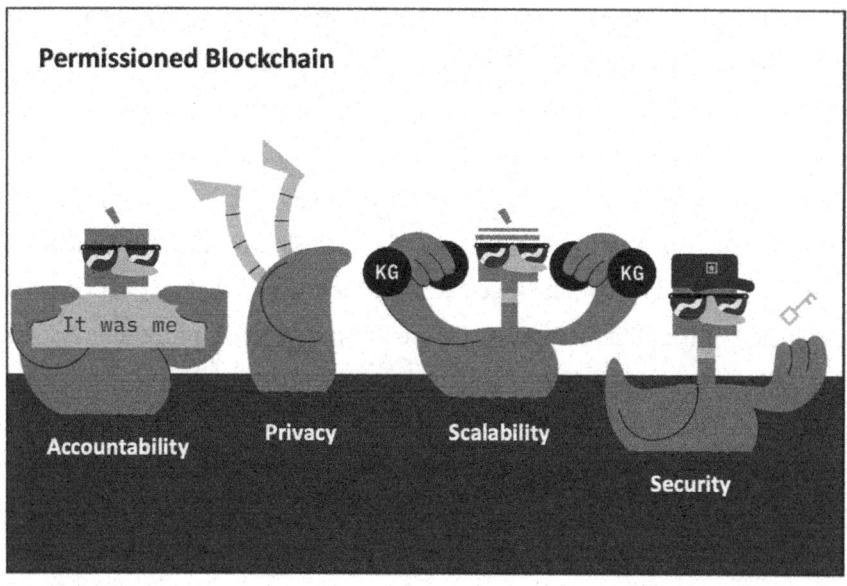

Chapter 1 – Figure 8 – Permissioned Blockchains exhibit these additional characteristics

Accountability

Accountability means network members are known (i.e., members are not pseudo anonymous) and identified by cryptographic membership keys with assigned access permissions by business role. Think about this as being similar to how a membership card works at your local health club. When you sign-up, you are issued credentials that give you access to the club with specific permissions to perform certain tasks at the club. For example, maybe your club-card allows you access to the Thursday night yoga classes. Enterprise blockchain member is similarly issued credentials (public/private keys) that give them access to reading or writing to the blockchain. Being known to the network is an important industry requirement. For example, without this, regulations like the EU General Data Protection Regulation (GDPR) would be near impossible to adhere to.

Privacy

While members are known to the network, transactions are only shared with those that have a need to know. Enterprise blockchain uses a variety of techniques to accomplish privacy, including peer to peer connections, channels, and zero knowledge proofs.

Scalability

Supporting an immense volume of transactions is critical to enterprise scenarios. Since transactions are not typically throttled in enterprise blockchains, as they are in networks like Bitcoin, they can get right down to business and perform. Of course, your transaction rates depend on several factors, including number of peers and complexity of the smart contract. With that said, transaction rates in the 1000s of transaction per second are certainly achievable.

Security

Enterprise blockchains are fault tolerant. With fault tolerant consensus algorithms, the network continues to operate even in the presence of bad actors or carelessness. An example of a fault tolerant consensus algorithm is the Byzantine Fault Tolerance algorithm, which we will study in an upcoming chapter.

- - -

Enterprise blockchains are often stereotyped as private networks, which I feel is an incorrect characterization. Access to an enterprise blockchain is set by "the group of administrators," who set the policy of how new members participate in the network. The visibility (e.g., public or private) of the network depends on how it is governed. So, it is true that enterprise blockchains are permissioned, but not necessarily private.

There is a fifth attribute that needs to be added to this definition, which is **finality**. Transactions in a blockchain network, especially if they are to be useful in a financial services context, need to be completed in real-time. Meaning, when a transaction updates an account balance… It is updated… now. Not later… And the network cannot "change its mind," which is also known as forking, which is sometimes done by permissionless networks.

There are two remaining points to mention before we conclude this chapter. The first is regarding the ducks. I hope you've enjoyed the visuals and analogy and they will help you appreciate the unique attributes of blockchains. Unfortunately, if you are not amused by the ducks, I'm afraid their usage in this book is not quite done. And the second point is, while public blockchain gets so much attention, I feel it is important to balance the conversation and share a little more on why I'm so excited about the role of enterprise blockchain applications in our world. So, here goes.

WHY DOES ENTERPRISE BLOCKCHAIN MATTER?

Because, today, no business operates in isolation. It should not be a stretch to think that multiple institutions could achieve more together than any single institution can alone.

By implementing new blockchain-based business processes that leverage the collective knowledge of the group, processes can become orders of magnitude more cost-efficient. And even more interesting, new cross organizational business processes can be created to open up new opportunity for these companies that were not possible before.

For example, the US Food and Drug Administration recently added food labeling regulations involving a requirement to notify the public of "sugar added" to food. How would a company, say producing protein bars, know for sure that the ingredients they are using contain sugar… and more importantly, prove it, if they are challenged. Operating on a trusted food blockchain, where ingredient suppliers recorded food information, the Protein Bar Company could easily show the provenance of each ingredient, from farm to "wrapper" to convenience store. This would certainly save time and money. But there's more. The same blockchain could now be used to save lives and allow the participating companies to trace back bad ingredients that may be causing food-borne illnesses. Given the food industry has many regulations to follow (much for our own safety and well-being), you can easily imagine how an enterprise blockchain is essential to making this scenario so. Specifically,

- Accountability - proving your institution is who you say you are to the FDA and other companies,

- Privacy - perhaps you don't want your competitor to know who you are buying your sugar from, and at what price,

- Scalability - there's lots and lots of food records,

- Security - must trust all information and information access must be resilient

With this example, I hope you can better appreciate blockchain with a more balanced understanding and now agree that not all blockchain applications involve crypto (not that there is anything wrong with that – a Seinfeld reference). But I could not in clear consciousness proceed without performing one more balancing act. The next section concludes this chapter with some inspiring examples of blockchain applications above and beyond Bitcoin.

BITCOIN IS ONLY ONE APPLICATION OF BLOCKCHAIN

I like to say that if there are one thousand applications of blockchain, Bitcoin is only one. In fact, the focus of this book is to also shine some light on the other nine-hundred and ninety-nine applications. As you will see, blockchain applications can go far beyond cryptocurrency and Bitcoin.

Now in all fairness, Bitcoin gets the credit for coining (no pun intended) the term 'blockchain.' The history of Bitcoin started with the invention and was implemented by the presumed pseudonymous Satoshi Nakamoto, who integrated many existing ideas from distributed computing, database, and the cryptography communities. Over the course of Bitcoin's history, it has undergone rapid growth to become a significant store of value both on- and-offline. From the mid-2010s, some businesses began accepting bitcoin in addition to traditional currencies.[6]

Bitcoin is an **application** that uses a specific variety of blockchain technology. Blockchain as a **technology** has many varieties and implementations. With its ability to create more transparency and fairness while also saving businesses time and money, the technology is impacting a variety of sectors in ways that range from how contracts are enforced to making government work more efficiently.[7]

Throughout this book I suggest we unhitch crypto from blockchain. They are related but different. In fact, as I write this book, crypto is going through its second significant downturn. The first occurred in 2018. While crypto will have its ups and downs, blockchain technology need not be put on the same rollercoaster. As you will read, permissioned blockchain technology is fueling a wide variety of novel applications that are changing everyday life for the better. All done independently from crypto.

The following list is just a small glimpse of the types of real-world blockchain applications for this pragmatic yet revolutionary technology. It's far from an exhaustive list, but they're already changing how we do business:

- Registries for land, vehicles, guns, etc.

- Medical data and clinical trial marketplace

- NFT marketplaces

- Cryptocurrency exchange

- Real estate management

- Music rights and royalties tracking

- Cross-border payments

- Identity management for people, places, and things

- Anti-money laundering tracking system

- Supply chain, logistics monitoring, and dispute management

- Voting

- Advertising insights

- Media and News authenticity

QUIZ TIME

As promised, here is the first round of quiz questions, which should be a piece of cake. And please be sure to get Question 4 right. You know I feel strongly about that one. Oh heck, the answer is FALSE.

1- Which is not a standard feature of a blockchain?

a. Shared ledger

b. Consensus mechanism

c. Support for Delete and Update commands

a. Immutability

2- Which terms most fit a permissionless blockchain?

    a.   Pseudo-anonymous administrators

    b.   Decentralized

    c.   Cryptocurrency as most common application

    d.   All of the above

3- Enterprise blockchains have these additional characteristics

    a.   Shared ledger, zero-knowledge proof, and BFT consensus

    b.   Accountability, privacy, scalability, and security

    c.   Decentralized, mining, public and permissionless

    d.   None of the above

4- Blockchain is Bitcoin?

    a.   True

    b.   False

UP NEXT

A pet peeve of many of my colleagues is when people carelessly use the term 'blockchain'. Sometimes loose terms like "the blockchain" and "blockchains" often overdramatize or even incorrectly portray the specific intent of blockchain's usage. For example, the term "the blockchain" incorrectly implies that there is a single blockchain network or technology, which is not true. So, for those colleagues that are sensitive to blockchain-term-abuse, I must apologize in advance for my indiscretions. However, when and if I use the term 'blockchain' loosely, I am almost always referring to blockchain technology in the broadest sense, as we will cover in detail in the chapters to follow.

- - -

Now that we have defined blockchain and have a better appreciation for the varieties of blockchain technologies and the applications therein, the next chapter will share three real blockchain stories that bring these concepts to life and will help you gain a better feel for how blockchain technology is being applied in the real world. You will also see in each of these stories that blockchain technology is truly being used for good. Read on to see what I mean.

From time to time, I will also share some photos from my blockchain adventures around the world. Here are a few that pertain to this chapter.

Chapter 1 – Figure 9 – Presenting at the Blockland (left) and
Think 2019 (right) where the "ducks" were unveiled

The Blockland Conference in Cleveland Ohio on December 4th, 2018, was the first time I introduced the blockchain ducks, before Shawn provided his iconic visuals. Also, I wore my Baker Mayfield jersey (see photo above).

There's a cool story behind that. The short version is: The coordinator of the conference had connections with the Cleveland Browns, and knew I was a fan of their new rookie quarterback. So, before my presentation, the GM of the Browns came on stage and gave me Baker's jersey. After the conference, I donated the jersey to a young man who was recuperating from cancer surgery, who had more reason to enjoy the jersey than me (also pictured in photo above).

To the right is a photo taken while I was presenting at IBM Think 2019 in San Francisco, California, February 14th, 2019. This was the first time I shared the duck analogy with Shawn's iconic ducks. Also, notice my bow tie. You'll see me sporting them from time to time.

Chapter 1 – Figure 10 – Blockchain Graphic Design by Shawn Lynch

The photos above feature more graphic designs from the talented Shawn Lynch. Shawn first introduced the iconic cartoon figures for my Think 2018 presentation, with blockchain ducks following at Think 2019. You will be seeing cartoon figures like these featured throughout this book.

Chapter 2 –
THREE BLOCKCHAIN
STORIES

Blockchain is changing everyday life for the better for people like you and me. These three stories are proof.

COVERED IN THIS CHAPTER

- The story of making food safe

- The story of eliminating identify theft

- The story of counterfeit prevention

CHANGING EVERYDAY LIFE FOR THE BETTER

How many of you believe that blockchain is changing everyday life for the better? Alright, perhaps you think it's too soon to make such a bold statement. Well, in this chapter, we are going to take a brief look at three examples (stories) that I am hopping will inspire you as much as they inspire me. I should add that each of these three stories is real and happening today.

It is so often the case when working in the information technology field that your collaborations and deliverables are happening deep inside the back-office of an enterprise that most of the world never gets to see or appreciate. These are the types of engagements that are hard to explain to your friends and/or grandfather when they ask you what you work on.

However, I've found my time working on blockchain quite different. Many of the blockchain-based solutions that I've had the privilege of working on are quite easy to explain and communicate to a broader, non-technical audience. The stories that follow certainly fit that profile. And what makes me really excited and proud to tell these stories is that these applications are actually making the world, and everyday life, better for folks like you and me.

REDUCING FOOD-BORNE ILLNESSES

Has this ever happened to you? You are rushing through LaGuardia Airport trying to make a flight. You are hungry and grab a sandwich before hopping on the plane. It's an hour into the flight and you are not feeling so well. Have you ever gotten sick after eating bad food? In 2006, a nationwide outbreak of E coli was linked to bagged spinach. It took regulators two weeks to conduct the trace back and determine the exact source of the outbreak.[8] During the two weeks, many people got sick, and one person died. Tons of good spinach was wrongfully wasted because we couldn't tell the good from the bad. Scientists have now shown that difficulties in finding the sources of contamination behind food poisoning cases are inevitable due to the increasing complexity of a global food traffic network, where food products are constantly crossing country borders, generating a worldwide network.[9]

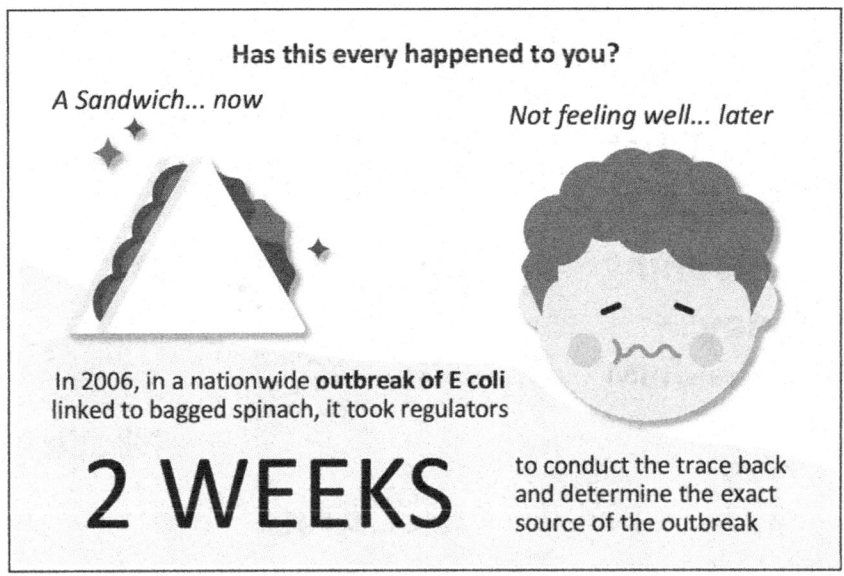

Chapter 2 – Figure 1 – Tracing back the source of food-borne illnesses can take weeks

Food Trust

The Food Trust Network consists of several major food companies, including Walmart, Unilever, and Nestlé. Convened by IBM, this network enables supply chain visibility across these members (and their ecosystem) to quickly pinpoint the sources of contamination. The network is already showing results, which reduce the impact of food recalls and limit the number of people who get sick or die from food-borne illnesses. Food Trust is a one-of-a-kind network that connects participants across a food supply-chain through a permissioned, permanent, and shared record of food system data. Not only has this network resulted in increased food safety and freshness, but it also unlocks supply chain efficiencies, minimizes waste, and enhances brand-reputation of the participants.

Chapter 2 – Figure 2 – The Food Trust network can instantly pinpoint the origin of food

Blockchain pinpoints the origin of food in seconds

With blockchain, network members can track provenance of ingredients as they travel from "farm to fork." Members can decide how they want suppliers to upload information to the shared ledger. In some cases, the information might come from spreadsheets maintained by a warehouse; for example, logging the arrival of a shipment of olive oil. Once on the blockchain, grocers and retailers working with the warehouse can access the information but can't alter or change it to cover up mistakes (you can flag mistakes, though, if need be).

For growers and producers, they might be uploading information from their phones or devices in the field, tagging bags of coffee beans as they're getting filled, or marking shipments of shrimp as they're brought in from the ocean. Everyone who the company has given access to their blockchain can access the information through a single portal run by Food Trust. Even end customers, like you and me, can see information about their food using that same data, though usually filtered through an app designed by the company.

International food companies Carrefour and Nestlé collaborated to add Mousline purée, a popular instant mashed potato mix available in France, into the Food Trust blockchain network. Ensuring the potatoes are grown in France, and of the highest quality, is important to French consumers, making this an ideal food to provide traceability. A barcode on the 520g packages can be scanned with a smart phone, giving consumers valuable information about the mix they're about to prepare, including the region where the potatoes were grown and varieties used, quality control in the Nestlé factory, where the product was made, and the places and dates of storage before it reached the grocer.

As you can see, the story of Food Trust goes a long way toward illustrating how blockchain can take a run at eliminating foodborne illnesses through trusted data sharing at scale. In fact, Walmart put this theory to the test by doing an experiment, which traced the origin of sliced mangos from a Walmart store back to the farm. They showed a radical improvement from the ~7 days it took to conduct the trace back using traditional methods down to 2.2 seconds using Food Trust. Frank Yiannas from Walmart says, "That's food traceability at the speed of thought," and I think you will agree that that's an inspiring story of blockchain changing everyday life.

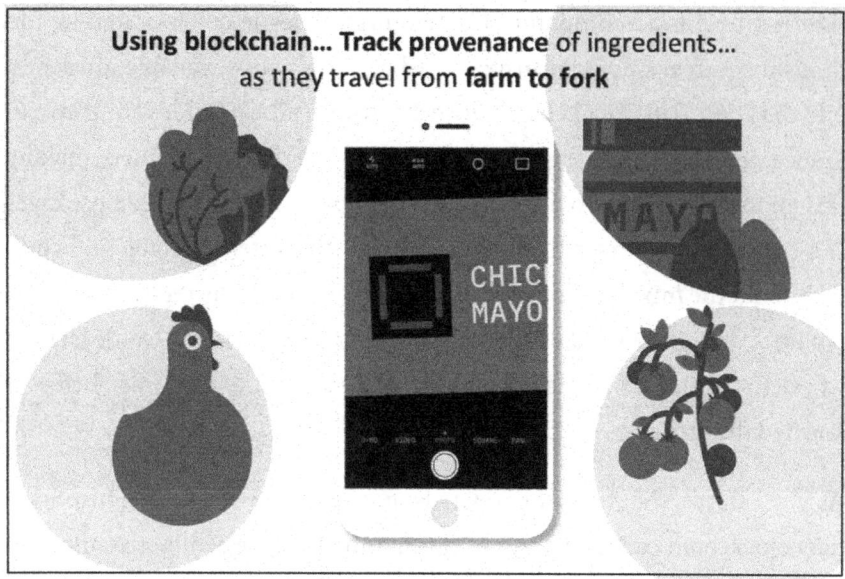

Chapter 2 – Figure 3 – Blockchain is used to track provenance of ingredients

ELIMINATING BIG DATA BREACHES

Has this ever happened to you? You are renting an apartment. The real estate company seemingly asks you to share information about every aspect of your life; where you live, your mother's maiden name, social security number, place of employment, and a credit statement from your bank. You repeat this process when you sign up for a new smart phone and visit the doctor for a checkup.

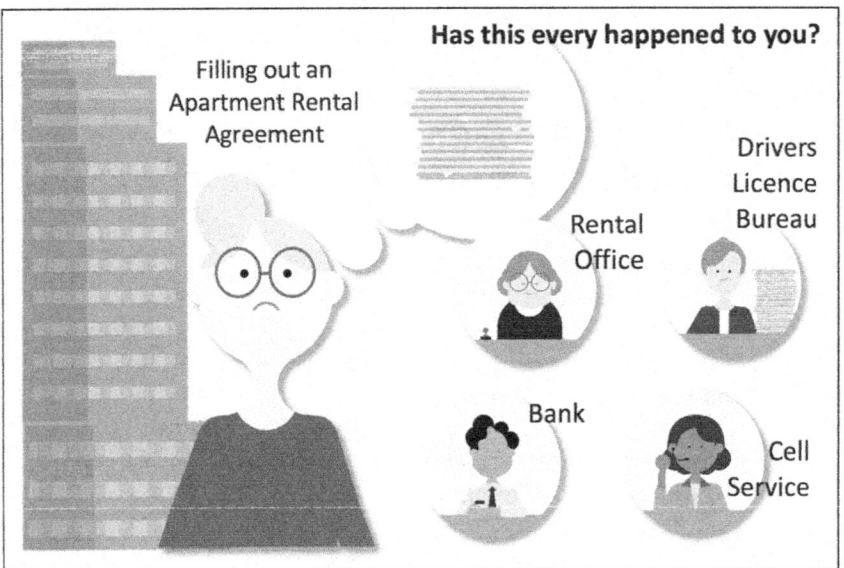

Chapter 2 – Figure 4 – Sharing too much personal information is a common affair

I don't know about you, but my digital life is a mess! I have bits and pieces of information, including user ids and passwords, scattered all over the "interweb." And then it happens. You get a notice from a major service provider that your data has been breached! Pretty scary and very frustrating.

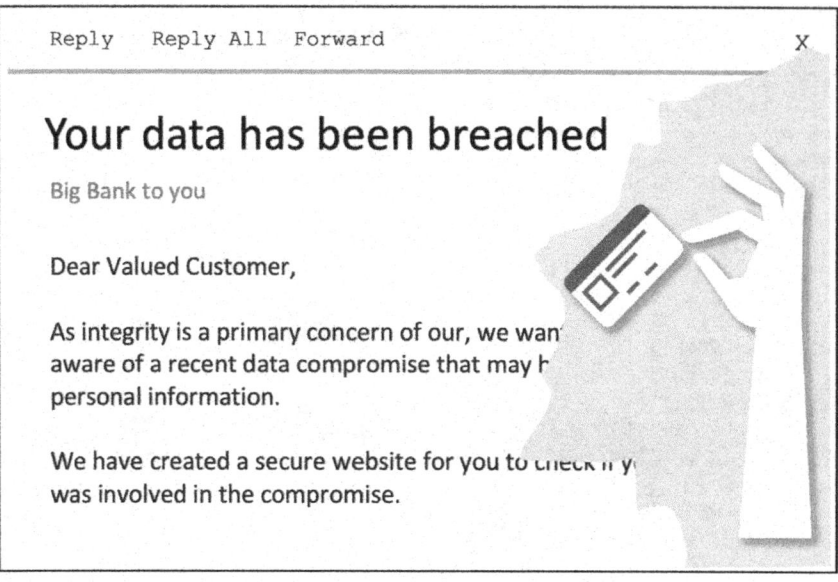

Chapter 2 – Figure 5 – Data breaches are all too frequent day-in-a-digital-life occurrences

In a 2021 Identity Fraud Study, released by Javelin Strategy & Research, a daunting new threat to consumers and businesses is revealed: identity fraud scams. Total combined fraud losses climbed to $56 billion in 2020. With traditional fraud, consumers often have no idea how their identities were stolen. With scams, they can often tell, upon reflection, the exact moment when they interacted with a criminal via email, phone, or text.

"The pandemic inspired a major shift in how criminals approach fraud," said John Buzzard, Lead Analyst, Fraud & Security with Javelin. "Identity fraud has evolved and now reflects the lengths criminals will take to directly target consumers in order to steal their personally identifiable information."[10]

Verified.Me

Verifying your identity can be an inconvenient task. Waiting in long lines to show several pieces of ID and answering multiple security questions are just a few of the hassles that come with verifying your identity to get access to the services and products you want. Not only is it inconvenient but how can you be assured that the personal data being shared is secured by the companies verifying you in case of a data breach?

The days of big data breaches are numbered with the emergence of the Verified.Me network convened by SecureKey. Verified.Me is a digital identity verification network developed in cooperation with seven of Canada's major financial institutions—BMO, CIBC, Desjardins, National Bank of Canada, RBC, Scotiabank, and TD. With the Verified.me smartphone app, you can finally take control of your digital identity attributes. The app provides a simple experience for signing up for (and signing onto) internet-based services. Acting as a digital rights management system for your identity, the app allows you to give permission to the real estate company to electronically ask the questions required to rent an apartment. You similarly give permission to trusted institutions (e.g., your banks, state/province Motor Vehicle Department, employer) to answer the real estate application questions.

Blockchain protects your personal information

With blockchain, the verification process takes place in real time and with unprecedented respect for your privacy. The solution is design such that there is no central database of identity information. The blockchain ledger is used as a digital rights management system, storing permissions and proofs by which the user grants institutions rights to access your identity information. Without a central "data honeypot," the attack surface of an identity breach is radically changed, making it very difficult for bad actors to walk away with a "big score." Blockchain also prevents your digital data from being tracked. Perhaps you don't want the real estate company to know which bank you do business with. By enabling "triple blind data exchanges," the data requester never knows who the provider is, the data provider knows not of the requester, and the network operator knows neither. Lastly, Blockchain enables only the necessary information to be exchanged.

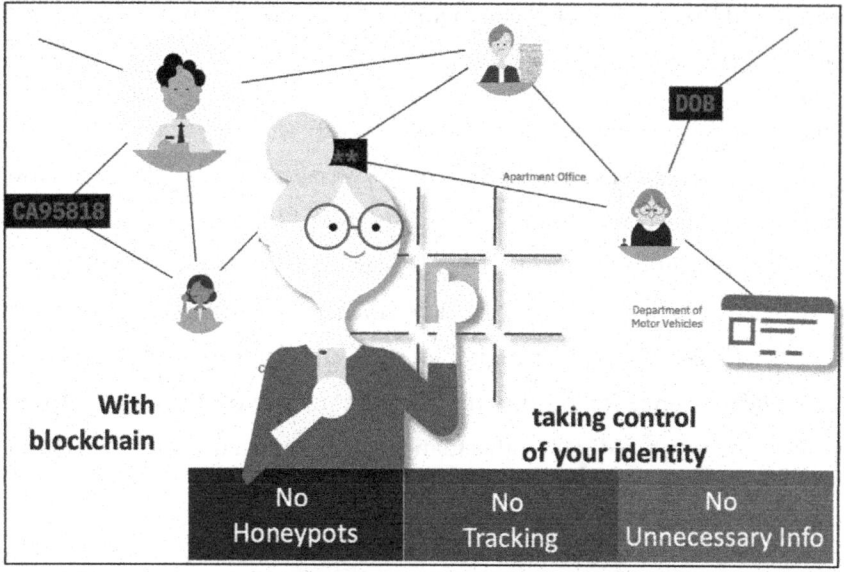

Chapter 2 – Figure 6 – Blockchain is used to eliminate big data "honeypots"

"In digital security, the less stray information floating around the better. The fewer companies storing your financial records, the less likely they'll be exposed in a breach. But though there are lots of ways to cut down on data sharing and retention, there are some things services just need to know, right?" asked Greg Wolfond, CEO of SecureKey.

Thanks to the cryptographic method known as "Zero-Knowledge Proofs" (ZKP), that's not always the case. Zero-knowledge techniques are mathematical methods used to verify things without sharing or revealing underlying data. Think of a payment app checking whether you have enough money in your bank account to complete a transaction without finding out anything else about your balance. Or an app confirming a password's validity without needing to directly process it. In this way, zero-knowledge proofs can help broker all sorts of sensitive agreements, transactions, and interactions in a more private and secure way.[11] We will go over ZKP in greater detail in a chapter to follow shortly. Stay tuned.

The National Institute of Standards and Technology (NIST), along with other privacy agencies view this approach as being best of breed for protecting user privacy. This is a big deal, and I think you would agree is another story of blockchain changing everyday life.

ANTI-COUNTERFEITING

Has this ever happened to you? This one actually happened to me. A friend, who had a headache, asked for an aspirin. I went into my "road-warrior aspirin-jar," that I keep in my laptop bag, and gave him one. Before he took it, he looked at it and asked, "What is this you're giving me?" It didn't look like any aspirin he had ever seen. I said, "I think it's an aspirin." He responded with a bit of terror in his voice, "You THINK?" The pill had a number on the side, which I quickly googled on my phone and realized it was a generic form of Tylenol. Phew! But not knowing the authenticity of medicine is a serious social issue that is considerably more widespread than my mislabeled generic Tylenol incident.

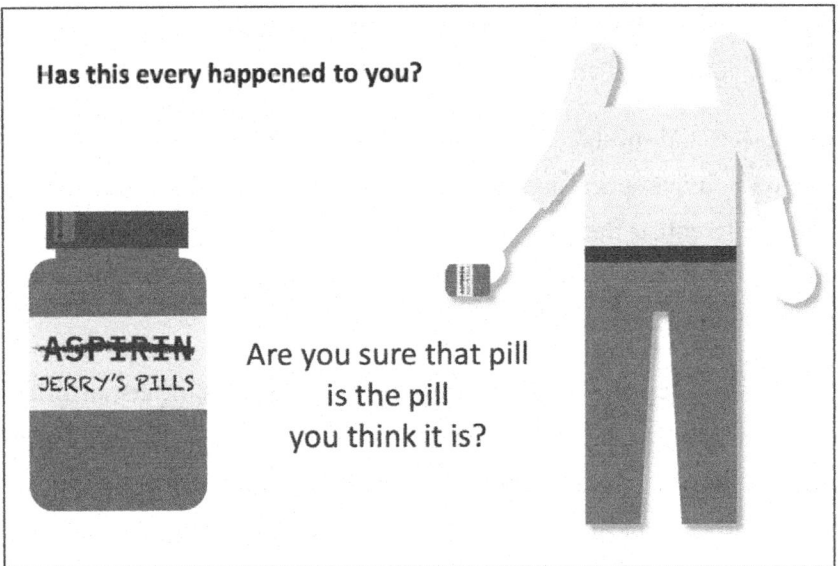

Chapter 2 – Figure 7 – How do you really know that item is authentic?

The World Health Organization estimates that 1 in 10 medical products circulating in low- and middle-income countries are either substandard or falsified, including pills, vaccines, and diagnostic kits. Examples include cough syrup for children that contained a powerful opioid and fake anti-malarial pills just made of potato and cornstarch.

"Substandard and falsified medicines particularly affect the most vulnerable communities," says Dr Tedros Adhanom Ghebreyesus, WHO Director-General. "Imagine a mother who gives up food or other basic needs to pay for her child's treatment, unaware that the medicines are substandard or falsified, and then that treatment causes her child to get even sicker. This is unacceptable. Countries have agreed on measures at the global level—it is time to translate them into tangible action."[12]

Crypto Anchor Verifier
IBM Research has developed IBM Crypto Anchor Verifier, a technology that brings innovations in AI and optical imaging together to help prove the identity and authenticity of objects. One of the first clients to try out this breakthrough was the Gemological Institute of America (GIA), to help them evaluate and grade diamonds.

The objects and substances that we buy, wear, eat or use every day all have their own unique optical patterns, sometimes undetectable by the human eye, which differentiate them from each other. These patterns can distinguish an organic ear of corn from a genetically modified one, or identify impurities in diamonds or aspirin pills, for example.

Optical characteristics can be measured using light spectrometers, an instrument used to measure properties of light, but they are quite bulky and expensive, limiting their utility. Use of optical analytics for object identification in real-world circumstances demands a more adaptable tool. IBM Research developed a powerful, portable optical analyzer, small enough to use with a cell phone camera.[13]

Blockchain can tell if it's a fake or legit

This innovative technology is a natural partner to blockchain technology. As we've just reviewed in the Food Trust story, interest in using blockchain to track physical goods is growing rapidly. IBM's Crypto Anchor Verifier solution can be used to capture the optical signature from an original, uncompromised item and subsequently record it on the blockchain, which can verify throughout the supply chain that the item hasn't been tampered with. As an example, Verifier can be used via the blockchain to ultimately confirm that the aspirin that you're taking for your headache is indeed the aspirin that the drug manufacturer shipped from its distribution center.

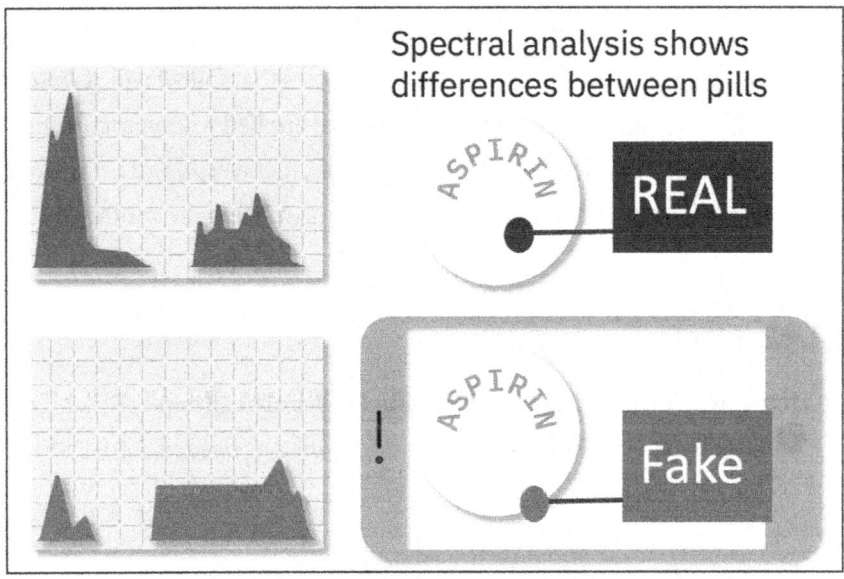

*Chapter 2 – Figure 8 – Digital profiles of legitimate goods can
be stored on a blockchain to detect counterfeits*

Verifier provides a lens attachment to a standard smartphone. The Verifier app leverages AI technology to perform light spectral analysis against a physical asset. It captures microscopic properties, viscosity, and other identifiers and produces a unique digital identifier for physical goods. When immutably placed on a blockchain, the "fingerprint" of that digital good can be checked again with Verifier by Customs, at a point of purchase, or checked right before you swallow your medication.

That's preventing counterfeiting with blockchain and it's changing everyday life.

- - -

I hope these stories illustrate that blockchain technology can be generously applied to problems being faced across all of today's industries. And while blockchain is best known as the technology that underpins the world's public and permissionless crypto networks, it also has a place to improve trust across participants in business networks. While these three stories happen to build on permissioned blockchain, the chapters that follow will certainly balance out the examples, covering blockchain applied to cryptocurrency, tokens, universal identity, decentralized finance, metaverse, and more. Two chapters down and a few more to go, but first it's time to take another quiz.

QUIZ TIME

1 - How does blockchain aid in reducing food-borne illness?

    a.   Food data can be exchanged and modified as needed across a food supply chain

    b.   Food data can be exchanged and not tampered with, creating trust and transparency of data across a food supply chain

    c.   Food companies can buy and sell food on mobile devices

    d.   None of the above

2 - How does blockchain protect personal identity?

    a.   No single "honeypot" of data where identity information is centrally kept

    b.   Makes it difficult to track data with triple blind data exchanges

    c.   Zero-Knowledge Proof means you only share the data that is needed. No more, no less.

    d.   All of the above

3 - What problem does Crypto Anchor Verifier solve?

    a.   Digitizes physical goods using light spectral analysis, tracked on a blockchain to eliminate counterfeiting

    b.   Taking photos of physical goods with your cell phone camera to buy and sell as NFTs

    c.   Eliminates identity theft by recording your fingerprint on a blockchain

    d.   Tracks food data on a food supply chain to eliminate food-borne illnesses

4 – Blockchain technology is an "equal opportunity employer."
It can be applied to use cases across all industries?

    a.   True

    b.   False

UP NEXT

For me, the stories covered in this chapter provide signs of the promise of blockchain becoming real and something that we all can be proud of. In the early 2000s, we started to see the first real signs of the Internet making the world a better place. It's energizing to see blockchain providing evidence in today's world that it is changing everyday life for the better.

I hope you now also see that blockchain can do much more than power cryptocurrency networks. As I alluded to in chapter one, there are thousands of applications that will benefit from blockchain technology, especially where trust and transparency are vital.

Given the stories, I thought it would be appropriate to share a few related photos from my photo roll to both Verified.Me and Food Trust.

Chapter 2 – Figure 9 – Jerry with Greg Wolfond and Frank Yiannas.

The photo (above, left) was taken with my friend Greg Wolfond, CEO of SecureKey and convener of the Verified.Me network, in San Antonio, Texas, in July 2017. The right photo is with Frank Yiannas, who was with Walmart and Food Trust, and now oversees food safety at the US Food and Drug Administration. This photo was taken when we both testified in front of Congress on February 14th, 2018. The photo that follows below is of Donna Dillenberger from IBM Research. In the photo Donna is demonstrating how Crypto Anchor Verifier can pinpoint real and counterfeit food products.

Chapter 2 – Figure 10 – My friend Donna Dillenberger demonstrating Crypto Anchor Verifier at Think 2019

- - -

While up to this point in the book, we have more than just hinted at how blockchain technology works, for the next chapter, it is time to roll up our sleeves as we will explore the inner workings of blockchain in detail.

Chapter 3 –
How Blockchain Works

Are you ready to create your own blockchain? Well, in this chapter we're going to do just that.

COVERED IN THIS CHAPTER

- Brief history of blockchain

- Blocks, chains, consensus, and distributed network by example

- Coding a blockchain

CODE-PHOBIA?

Are you ready to create your own blockchain? Well in this chapter, we're going to do just that. But if coding is not your thing, that's okay. I promise this chapter will still be quite informative and the code example will be easy to follow and illustrate the concepts we cover earlier in this chapter. And for those of you who don't have "code-phobia," I hope you appreciate that while the coding example is just a simple skeleton of a real blockchain, it will act as a picture and "speak one thousand words" to further solidify these concepts. But, before we build a blockchain, I want to revisit some of the concepts that were introduced in the Blockchain-Ducks Chapter, but this time, go into more detail on the key concepts that make blockchains work.

A BRIEF HISTORY

To understand how blockchain works, it helps to understand some of the history surrounding the technology. The technique of cryptographically "chaining blocks" containing information was originally described in 1991 by Stuart Haber and W. Scott Stornetta. Their paper, *"How to time-stamp a digital document,"* details a software system that behaves as a "notary public," time stamping digital documents to prevent backdating and tampering with the document's content. The paper abstract reads as follows:

The prospect of a world in which all text, audio, picture, and video documents are in digital form on easily modifiable media raises the issue of how to certify when a document was created or last changed. The problem is to time stamp the data, not the medium. We propose computationally practical procedures for digital time stamping of such documents so that it is infeasible for a user either to backdate or to forward-date his document, even with the collusion of a time stamping service. Our procedures maintain complete privacy of the documents themselves and require no record-keeping by the time stamping service. [14]

Haber and Stornetta's work, however, went mostly unnoticed until the first decentralized blockchain was conceptualized by a person (or group of people) known as Satoshi Nakamoto in 2008. Nakamoto is widely cited as the inventor of the digital cryptocurrency Bitcoin. Nakamoto's disappearance is at least as brilliant as the technology he created, which is why we have dedicated an entire supplementary chapter to the entity known as Satoshi Nakamoto. Satoshi's paper titled "Bitcoin: A Peer-to-Peer Electronic Cash System" foresaw the need for a peer-to-peer online payment system that is self-governing, secure, and limited in quantity. The Bitcoin network was launched on January 3, 2009, with each bitcoin priced at $0.0008. The paper abstract reads as follows:

A purely peer-to-peer version of electronic cash would allow online payments to be sent directly from one party to another without the burdens of going through a financial institution. Digital signatures provide part of the solution, but the main benefits are lost if a trusted third party is still required to prevent double-spending. We propose a solution to the double-spending problem using a peer-to-peer network. The network time stamps transactions by hashing them into an ongoing chain of hash-based proof-of-work, forming a record that cannot be changed without redoing the proof-of-work. The longest chain not only serves as proof of the sequence of events witnessed, but proof that it came from the largest pool of CPU power. As long as honest nodes control the most CPU power on the network, they can generate the longest chain and outpace any attackers. The network itself requires minimal structure. Messages are broadcast on a best effort basis, and nodes can leave and rejoin the network at will, accepting the longest proof-of-work chain as proof of what happened while they were gone.[15]

The blockchain technology powering Satoshi's Bitcoin is a great place to start our study of how blockchain works because it possesses many of the most basic features of a blockchain. Bitcoin is also a simpler place to start because it has yet to be complicated by advanced concepts like, pluggable consensus and smart contracts (which we will also study later in this book) that exist in some of the newer blockchain technology implementations.

In chapter one, we discussed the property of immutability, which is one of the most basic and interesting aspects of all blockchain technology.

ACHIEVING IMMUTABILITY

The abstracts listed above both allude to this property of immutability as the key breakthrough of blockchain and forms the foundation of trust. So how does immutability work? To understand this, we need to understand what we mean by "blocks" and then "chains." Here goes.

Blocks and chains

Let's start with a **block**. A block is a place in a blockchain where information is stored and encrypted. As seen in the following figure, a block contains **data, the hash of the block,** and the **hash of the previous block**.

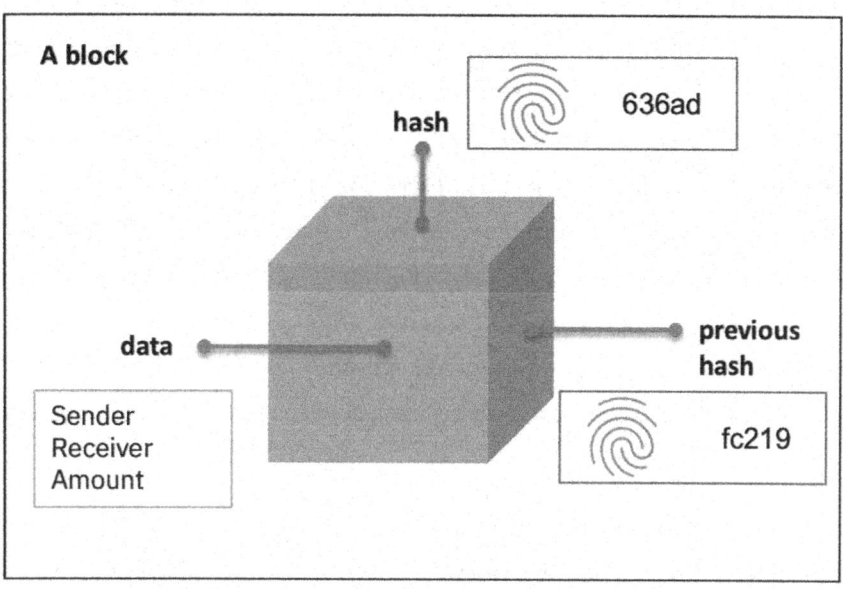

Chapter 3 – Figure 1 – A Block; containing data, hash, and previous hash

The structure of the **data** stored inside a block depends on the type of block-chain. The Bitcoin blockchain, for example, stores details related to transaction such as **the sender's address, the receiver's address, and an amount**.

A block also has a cryptographic **hash** that is used as the primary identifier of a block. Hashing requires processing the data from a block through a mathematical function, which outputs a unique number. Therefore, no two blocks have the same hash value; hence they can be thought of to be as unique as your fingerprint. The hash is calculated upon creation of the block. Hashes are very useful for detecting changes to blocks as we will see in the upcoming example.

The third element of a block is the **hash of the previous block**. This effectively creates a chain of blocks and it's the chaining technique that makes a blockchain so secure.

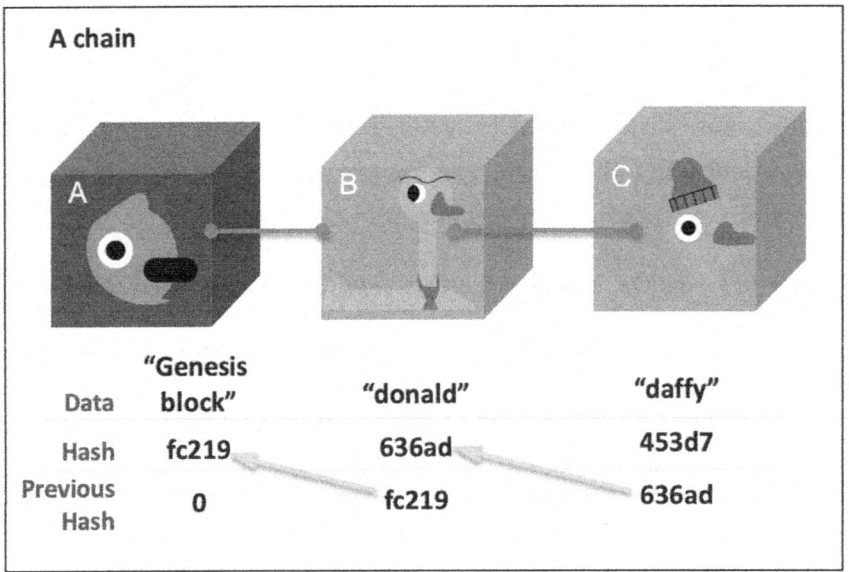

Chapter 3 - Figure 2 - A Chain of cryptographically linked Blocks

The above example is of a simple chain of three blocks. As described, each block has a cryptographic hash and a hash of the previous block. Specifically, the figure shows:

- block C, linking to block B

- block B, linking to block A

Genesis block

Now, block A is a bit special because it is the first block in the chain and there isn't a previous block to link to. The block is referred to as the **genesis block** (and sometimes also called **block 0**) and is effectively the ancestor that every other block can trace its lineage back to since every block references the one preceding it.

Fun fact (that will not be on the quiz following this chapter): According to Bitcoin.it, the hash of the Bitcoin genesis block is:

*0x736B6E616220726F662074756F6C69616220646E6F63657320666F-
206B6E697262206E6F20726F6C6C65636E61684320393030322F6E614A2
F33302073656D695420656854*

A Bitcoin enthusiast found an interesting Easter Egg in the Genesis Block.[16] It's a comment describing a message Satoshi put in. When translated into Latin characters, it came with this interesting sentence:

"sknab rof tuoliab dnoces fo knirb no rollecnahC 9002/naJ/30 semiT ehT."

Can you guess what it says? You need to read the sentence backwards. It says:

"The Times 03/Jan/2009 Chancellor on brink of second bailout for banks."

The functions of the genesis block differentiate it from the rest of the blocks because it is the root of the whole chain, and without it, the system would not work. Therefore, every blockchain has its own genesis block no matter what type of blockchain it is. Genesis blocks also ensure the operation and communication between two nodes in a blockchain network. We will cover the networking aspects of blockchain in an upcoming section.

Back to our example. Let's take a closer look at what would happen if we tampered with one of these blocks by changing the data value found in block B. Tampering with block B causes the hash of the block to be recalculated, resulting in a new hash value. This simple act "breaks the chain" because block C's previous block link to block B is no longer valid. In fact, all blocks created after C (let's say, D, E, and F) are also made invalid because their previous block links are now pointing to invalid blocks.

Consensus

Altering the data of a single block will make all following blocks invalid. But using hashes is not enough to prevent tampering. Computers these days are very fast and can calculate hundreds of thousands of hashes per second. You could effectively tamper with the block and recalculate all the hashes of other blocks to make your blockchain valid again. To mitigate this, the Bitcoin blockchain inserts a process called **Proof of work** (POW) as a mechanism to slow down (or throttle) the creation of new blocks as well as validate the integrity of the chain (consensus). POW requires **miners** (i.e., nodes on a blockchain network) to solve a complex problem (i.e., hashing) that expends computational power (i.e., work) to achieve consensus and prevent bad actors from overtaking the network.

In Bitcoin's case, it takes about 10 minutes to calculate the required proof of work and add a new block to the chain. This additional step makes it very hard to tamper with the blocks because if you tamper with one block, you'll need to recalculate the POW for all the following blocks (each taking an additional 10 minutes to calculate). The security of the Bitcoin blockchain comes from its creative use of hashing and the POW mechanism.

POW is just one consensus approach used by popular blockchain technologies. POW is an energy-intensive process that can have trouble scaling to accommodate large numbers of transactions that can be generated by the growing variety of blockchain applications. **Proof of stake** (POS) is a popular alternative consensus mechanism, where only select nodes in the network are called on to validate transactions.

While POW and POS are popular consensus mechanisms for permissionless blockchain, where the participants are anonymous, there are several interesting and unique consensus algorithms used by permissioned blockchains, where the network participants are known. Consensus algorithms include practical **byzantine fault tolerance** (pBFT) and Reliable, Replicated, Redundant, And Fault-Tolerant **(RAFT)**. We will see a more detailed description of these types of algorithms in Chapter 6.

Distributed Ledger

There is one more way that blockchains secure themselves through a distributed versus centralized approach to data management. For example, the blockchain technology employed by Bitcoin uses a peer-to-peer network, which is open for anyone to join anonymously. When a participant joins this network (by adding a new network node), they get a full copy of the blockchain ledger. With a complete ledger, the node can add new blocks and verify the chain is still in order.

However, it is important to note that in a distributed blockchain network, everyone "has control," but no single node is "in control." Blocks are added based on a collaborative consensus process. When a node creates a new block, that block is first sent to the nodes on the network. Each node, using their exact copy of the ledger, then verifies the block to make sure that it hasn't been tampered with. If all the hashes check out, each node adds this block to their own blockchain. Therefore, all the nodes in the network consent as to which blocks are valid and which aren't. Blocks that are tampered with will be rejected by other nodes in the network.

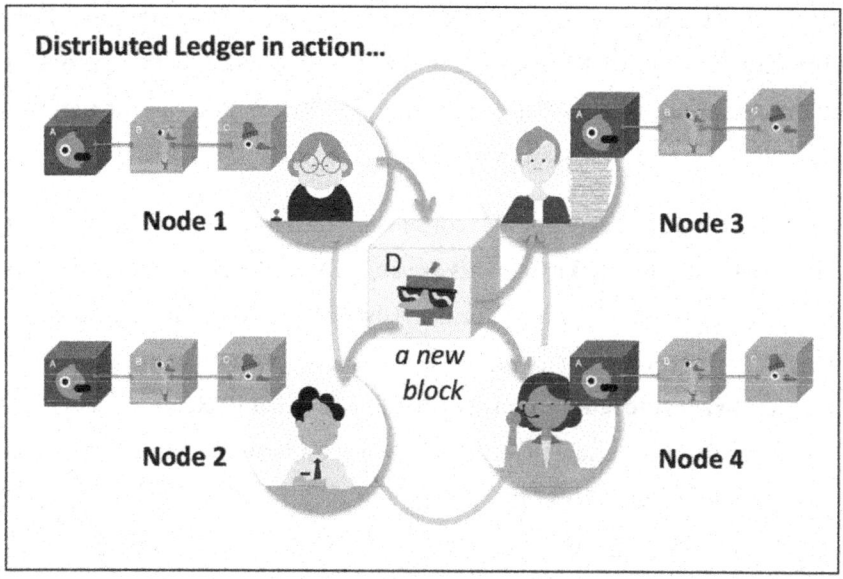

Chapter 3 – Figure 3 – A distributed blockchain network consenting to add a block

To maliciously (or accidentally) tamper with the blockchain, you'll need to tamper with all the blocks on the chain. Which includes rerunning the proof of work for each and every block and taking control of more than 50% of the nodes in the peer-to-peer network. Then and only then will your tampered block become accepted by everyone. This is highly improbable and why blockchains claim to be immutable and tamper resistant.

CODING A SIMPLE BLOCKCHAIN

In this section, we are going to write a very simple blockchain in the JavaScript language and give it a test-drive on node.js.[17] This exercise is designed to build just enough of a blockchain to further illustrate the concepts described above, including immutability, blocks, hashes, chains, and validating the chain's integrity. The total source code is 99 lines including comments and blank lines. We call our blockchain 'Duckchain,' for obvious reasons.

Let's get started.

Defining a block

We'll start by defining the structure of a **block** object. The code snippet below defines a *class* called **Block**, which has four properties: **index, data, hash, and previousHash**. The Block class has a *constructor* method that is invoked when a new Block is created and accepts three input parameters that set that Block's values for index, data, and previousHash. However, the hash property is not passed in as a parameter, because it needs to be calculated (see line 12).

```
 3   // This class defines a Block object destined for a Blockchain
 4   class Block {
 5
 6       // method that creates a new Block
 7       constructor(index, data, previousHash) {
 8
 9           // A Block on the "Duckchain" contains these properties
10           this.index = index;            // position of block in the chain
11           this.data  = data;             // data stored in the block
12           this.hash  = this.calculateHash();  // calculate a hash for this block
13           this.previousHash= previousHash;    // the hash of the previous block
14       }
15   }
```

Calculating the block's hash

A *hash* algorithm is a function that can be used to map data of arbitrary size to fixed-size values. Before we continue with our coding example, let's take a quick detour to look at a simple hashing algorithm to help crystallize this important cryptography aspect of blockchain. Imagine a hash algorithm, let's call it the ABC Hash, which takes as input a text string (e.g., "hello") and iterates through each letter of the string, adding up the letter's positional value in the alphabet (e.g., a=1, b=2, z=26). The resulting sum of all letters is the hash-value.

Input	ABC Hashing algorithm	Hash-value
hello	8 + 5 + 12 + 12 + 15	52

Hence in the example above, if we calculate the ABC hash of the word "hello," the resulting hash value would be "52." Hashing algorithms are considered efficient and secure because they are one-way functions and are non-reversible. For example, it would be nearly impossible to reverse the function by taking "52" as input and somehow producing "hello" as its sole output.

By default, JavaScript does not include cryptographic hashing functions. But this is not a problem because a complete library containing multiple hashing algorithms (except our fancy ABC hashing function) is both free and easy to install using the **node package manager** (npm)[18] using the following command.

npm install --save crypto-js

Before we look at how the *calculateHash* method is defined, we will need to import an appropriate hash function from the crypto-js library just installed. We will go with the popular Secure Hash Algorithm called SHA-256. I will not go into too much detail on what SHA-256 is, but I will just say to those crypto-aficionados that this is the go-to for most blockchain technologies because of its use of a 256-bit key, which is much more secure than other common hashing algorithms with an unlikely possibility of having the same hash value repeated (collisions).[19] All we need to do is add the following line to the top of our source file (as seen in line 1 in the code snippet below) and we have access to this powerful cryptographic hashing algorithm.

Now we are ready to add the calculateHash method to the Block class which starts at line 16 in the source code that follows.

```
 1   const SHA256 = require('crypto-js/sha256');
```

```
16   // method that calculates the hash of a block
17   // by applying a SHA256 hash function across block properties
18   calculateHash() {
19       let newHash = SHA256(this.index +                  // include the index
20                       JSON.stringify(this.data) +  // all the data
21                       this.previousHash);          // and the previous hash
22       return(newHash.toString());                  // returns a 64-character hash
23   }
```

Much of the "heavy-lifting" of this method is handled by the SHA256 function. We provide all the properties of the Block including the index, data (converted to a string), and the previous hash as input into the SHA256 function. We store the hash value result in a temporary variable (I like to do this to make the code a bit easier to read and fit on a page) and convert it to a string before returning it as the result.

Please appreciate that when we talk about the importance of the use of cryptography in blockchain, it really comes down to this sort of hashing. This completes the coding of the Block class. Pretty simple so far. Right? Now it's time to create the blockchain.

Defining a chain

As seen below, the **Blockchain** class is a little bit more involved than the Block class, so we will examine it in bite-sized snippets. The *constructor* method creates a new instance of a blockchain object, which comprises an array of Block objects called 'chain.'

```
26    // This class defines a blockchain object
27    class Blockchain {
28
29        // method that creates a new blockchain
30        // initializes the blockchain as array of blocks
31        constructor() {
32            this.chain = [this.createGenesisBlock()]; // 'chain' is the blockchain
33        }                                             // starting with genesis block
34
35        // method that creates block 0 aka the genesis block
36        createGenesisBlock() {
37            let genesisData = "Genesis Block Quack Quack!";
38            let genesisBlock = new Block(0,// the index is zero because its block 0
39                                genesisData,  // data for the genesis block
40                                "0");         // there is no previous block
41            return(genesisBlock);
42        }
43    }
```

Remember our discussion of the genesis block in an earlier section of this chapter? Well, here's where it comes into play. When the constructor method defines the chain array, it initiates it with a genesis block. Which is where the role of the next method, called **createGenesisBlock,** comes in. As you can see above, this method creates a new Block object, and initializes it with Block properties. I picked an arbitrary (yet appropriate) string for the data field. The index field is set to 0. Now you might get why the genesis block is also referred to as block zero. Before now, I really haven't said much about the **index** property of the Block object. This is where it comes in. The index is the index into the chain array and allows direct access to any Block in the blockchain, as we will see in the method descriptions to follow.

Adding blocks to the chain

The next code snippet includes a set of "helper" methods that add Block objects to the Blockchain object.

```
44      // fetches the last block added to the chain
45      getLatestBlock() {
46        let last = this.chain.length - 1; // calculate last item in the change
47        let lastBlock = this.chain[last]; // get the last block
48        return(lastBlock);                // return the last block
49      }
50
51      // adds a new block on to the chain
52      addBlock(newBlock) {
53          // fetch the hash of last block
54          newBlock.previousHash = this.getLatestBlock().hash;
55          // calc has of this new block
56          newBlock.hash = newBlock.calculateHash();
57          // add the new block to chain
58          this.chain.push(newBlock);
59      }
60  }
```

The *getLatestBlock* method does just that, returning the last block that was added to the blockchain. There may be many interesting uses of this method, however, it is used explicitly by the next method in the above snippet, called *addBlock*. This is an important method because it is how we push new Block objects on the Blockchain. The method takes a **newBlock** parameter, which doesn't yet have either of the hash fields populated. The method gets the hash from the last Block (using the aforementioned getLatestBlock helper method) and sets the **previousHash**. Now that the big three (index, data, and previousHash) have been set, the hash of this new block can be calculated. We finish by pushing the new Block on the chain.

Checking the integrity of the chain

Immutable means 'unchanged over time' or 'unable to be mutated.' Immutability is a defining feature of a blockchain as once created, blocks can't be changed without invalidating the integrity of the chain. The following important code snippet tests the validity of our blockchain and is called *isChainValid*.

```
62   // this method checks the integrity of the chain by checking the validity
63   // of the blocks hash and the current and previous block's hashes match
64   // return true if all checks out, otherwise false
65   isChainValid() {
66       // iterate through all block on the chain
67       for (let i=1; i < this.chain.length; i++) {
68           const currentBlock  = this.chain[i];      // get the current block
69           const previousBlock = this.chain[i - 1]; // get the previous block
70
71           // check hash is correct by recalculating
72           if (currentBlock.hash !== currentBlock.calculateHash()) {
73               console.log('Invalid hash - Block #' + currentBlock.index);
74               return false;   // nope - the block has been tampered with
75           }
76
77           // check previous blocks hasn't changed too
78           if (currentBlock.previousHash !== previousBlock.hash) {
79               console.log('Invalid previousHash - Block #' + currentBlock.index);
80               return false;   // nope - the block has been tampered with
81           }
82       }
83       return true;  // we made it... everything checks out!
84   }
85 }
```

This method inspects the current state of the entire Blockchain object and returns true if it's valid and false if not. As illustrated in the code above, the validation process involves iterating through each Block object listed in the **chain** array. This is achieved by the **for loop,** which starts at the 1st index and loops until it hits the last index. Notice that we do not start at index 0, which is the genesis block, which can be skipped.

To be valid, a Block must pass two tests. The first test involves recalculating the Block's hash and comparing it to the hash value currently set for the Block. The second test is to check that the hash of this Block's previous hash value is actually equal to the previous Block's hash value.

If, at any time, one of these conditions fails, the entire chain is considered invalid, and the function returns false. And if all checks out, we have a chain with integrity and return true.

Running and testing the Duckchain

With our simple blockchain complete, it's time to give it a test run. The following code snippet makes a Duckchain come alive.

```
87   // create and test the Duckchain
88   let duckchain = new Blockchain(); // Tada! duckchain is live on my laptop
89   duckchain.addBlock(new Block(1, "donald"));  // add the donald block
90   duckchain.addBlock(new Block(1, "daffy" ));  // add the daffy block
91   // print out the full duckchain
92   console.log(JSON.stringify(duckchain, null, 4));
93   // print out if chain is valid or not
94   console.log('Is the Duckchain valid? ' + duckchain.isChainValid());
95
96   // Tamper with the chain and test validity again
97   duckchain.chain[1].data = "howard";
98   duckchain.chain[1].hash = duckchain.chain[1].calculateHash();
99   console.log('Is the Duckchain valid? ' + duckchain.isChainValid());
```

We start by creating a new instance of a Blockchain object and assigning it to
the **duckchain** variable. Duckchain is now officially alive! We then proceed
to add a few Blocks to the Duckchain. Each Block has a data field of a popular
duck's name, creating one for Donald and Daffy. Now that we have a populated blockchain, I can print out (log) its content to my terminal's console
and also print a message regarding the validity of the Duckchain.

The test code then tries to tamper with the first block in the Duckchain,
changing the data field from Donald to Howard. We try to mask this devious
task by also recalculating the hash for this block thinking that perhaps no
one will notice. The test program then concludes by rechecking the integrity of the chain and prints out another message regarding the validity of the
Duckchain. Here are the results when we run the test code.

```
{
    "chain": [
        {
            "index": 0,
            "data": "Genesis Block Quack Quack!",
            "hash": "fc2193068c46fa6568886afcf5835030
                     a1db99abc1faad2205f864e191f0c839",
            "previousHash": "0"
        },
        {
            "index": 1,
            "data": "donald",
            "hash": "0b72f7611b4ad7c8c82dc8cc351bd7ec
                     d0fe5ca58629087b9a89fe2144fd1f8d",
            "previousHash":
                    "fc2193068c46fa6568886afcf5835030
                     a1db99abc1faad2205f864e191f0c839",
        },
        {
            "index": 2,
            "data": "daffy",
            "hash": "ac76324eabe385711b526e40a2ca4b6d
                     8f7b2fd93c5b5fd90880eac9ef7169ba",
            "previousHash":
                    "0b72f7611b4ad7c8c82dc8cc351bd7ec
                     d0fe5ca58629087b9a89fe2144fd1f8d",
        }
    ]
}
Is the Duckchain valid? true
Invalid previousHash - Block #2
Is the Duckchain valid? false
```

Looking at the results of our test above, we can first take note of the hashes for each entry in the Duckchain and how the hash and previousHash values line up correctly. At this point, we have yet to tamper with any Blocks, hence we see the message "Is the Duckchain valid?" with a value of true. However, we then see the value change in the last log message to false because our tampering was correctly detected.

So much more...

Congratulations! We've accomplished our mission in coding a trivial blockchain in under 100 lines of JavaScript. With this, we take away a more vivid appreciation of the relationship between a block and how it can be cryptographically chained to other blocks such that if blocks are tampered with, the chain becomes invalid.

From here, we can certainly imagine the possibilities of how we might extend this simple Duckchain into a more functional blockchain. For example, we can explore how the ledger might be distributed and synchronized. Once distributed, we can explore different consensus algorithms including proof of work and perhaps other algorithms that make bold assumptions based on the trusted identity of the participants running nodes. Distributing the ledger to multiple network participants introduces specific optimization, like **Merkle trees** (fully explained in the next chapter), which allow the ledger to be verified without rehashing the entire chain each time. And while we are at it, we can make the addition of **smart contracts**. Of course, those additions would add thousands of lines of code to this exercise, as well as exponential complexity, and we would perhaps want to reconsider our language choices of JavaScript and node.js. However, I believe we built enough to gain an appreciation of how a blockchain works as well as the art of the possible features beyond the ones covered here. And fear not, some of these advanced topics are covered, with coding examples, in the chapters to follow.

QUIZ TIME

1 – The technique of cryptographically chaining information including time stamped digital documents to prevent backdating and tampering was first described in this paper.

    a. "How to time-stamp a digital document" by Haber and Stornetta, 1991

    b. "Bitcoin: A Peer-to-Peer Electronic Cash System" by Satoshi Nakamoto, 2008

    c. "Beyond Bitcoin: Emerging Applications for Blockchain Technology", Testimony by Gennaro Cuomo, 2018

    d. None of the above

2 – Altering the data of a single block will make all following blocks invalid

    a.   True

    b.   False

3 – A cryptography algorithm preferred by many blockchain to hash blocks

    a.   ABC Hash

    b.   SHA-256

    c.   MD-6

    d.   All of the above

4 – What characteristic(s) best describes a genesis block?

    a.   Block 0 – i.e., first block in the blockchain

    b.   Only block on the chain that does not have a "previous hash" value

    c.   The ancestor that every other block can trace its lineage back to since every block references the one preceding it

    d.   All of the above

5 – Think Blockchain Labs – Chapter 3 Assignment

Go to the Think Blockchain Labs Github repository.

See the README.md file and search for the Chapter 3 assignment details, but the gist of the assignment is to:

    a.   Run the simple Duckchain sample

b. Enhance the Duckchain sample by modifying the data field in the Block object to approximate a Bitcoin transition, such that the data field includes a sender, receiver, and amount.

c. Implement a send method that sends some number of coins to a receiver.

UP NEXT

Many of the concepts covered in this chapter first came to my attention while walking around and talking to developers during blockchain hackathons. This photo was taken at my first blockchain hackathon in New York City in 2016. In the photo are some of Hyperledger's best hackers including (left to right) Gari Singh, Chris Ferris, and John Wolpert.

Chapter 3 – Figure 4 – My first blockchain hackathon

- - -

In the next few chapters, we will continue to explore the inner workings of blockchain technology by looking at three blockchain pioneering use cases. For each pioneer, we examine the three intimately-related topics that uniquely make the pioneer a legend in the blockchain world. The topics are:

- The Use Case

- Pioneering Network/Technology

- The Breakthrough Concept

Specifically in the next chapter, the pioneering network is Bitcoin, the use case is Cryptocurrency, and Mining is its breakthrough concept. Ready, set, go!

Chapter 4 -
CRYPTOCURRENCY, BITCOIN, AND MINING

A purely peer-to-peer version of electronic cash would allow online payments to be sent directly from one party to another without the burdens of going through a financial institution – Satoshi Nakamoto

COVERED IN THIS CHAPTER

- The evolution from fiat currency to crypto

- How the Bitcoin application creates an economy around a digital currency

- The anatomy of a Bitcoin block and the process of mining it

- A coding example illustrating the mining process

CRYPTO

Welcome to our first chapter that deep dives into a blockchain use case. What better application to study than arguably the first successful blockchain application, which is cryptocurrency as seen through the lens of Bitcoin. This chapter builds on the prior chapter, where we discussed how blockchains work at a high-level.

We'll start by refreshing our memories by tracking the history of currency, money, digital currency to cryptocurrency and Bitcoin. We then explore a few advanced topics in blockchain that enable Bitcoin to securely operate at a global level. And I couldn't resist adding another coding exercise to help illustrate the concept of Bitcoin mining.

Cryptocurrency, sometimes also simply called crypto, is a **digital payment system** that doesn't rely on banks to verify transactions. It's a peer-to-peer system that can enable anyone anywhere to send and receive payments. Cryptocurrency received its name because it uses encryption to verify transactions. When you transfer cryptocurrency funds, the transactions are recorded in a public ledger.[20] There are thousands of cryptocurrencies. Some of the best known include:

- **Bitcoin:** The first and most traded cryptocurrency is the prime subject of this very chapter.

- **Litecoin:** This currency is most like Bitcoin but has moved more quickly to develop new innovations, including faster payments and processes to allow more transactions.

- **Ethereum:** This is a blockchain platform with its own cryptocurrency called Ether (ETH). It is the most popular cryptocurrency after Bitcoin and pioneered an implementation of smart contracts, which has fueled new uses of blockchain including NFTs. We will cover this in the next chapter.

Before we take a giant step into this fascinating use of blockchain technology, the next section will set the stage by outlining the evolution of currency to crypto.

FROM BARTERING TO FIAT CURRENCY

Currency is a medium of exchange for goods and services. In short, it's **money** in the form of paper or coins, usually issued by a government and generally accepted at its face value as a method of payment.[21]

What exactly is money? At its core, money represents **value**. For example, if I do some work for you, and you give me money in exchange for the value I gave you; I can then use that money to get something of value from someone else in the future. Throughout history, value has taken many forms and people have used different materials to represent money. Salt, wheat, seashells, ducks, and of course, gold have all been used as a medium of currency exchange.[22]

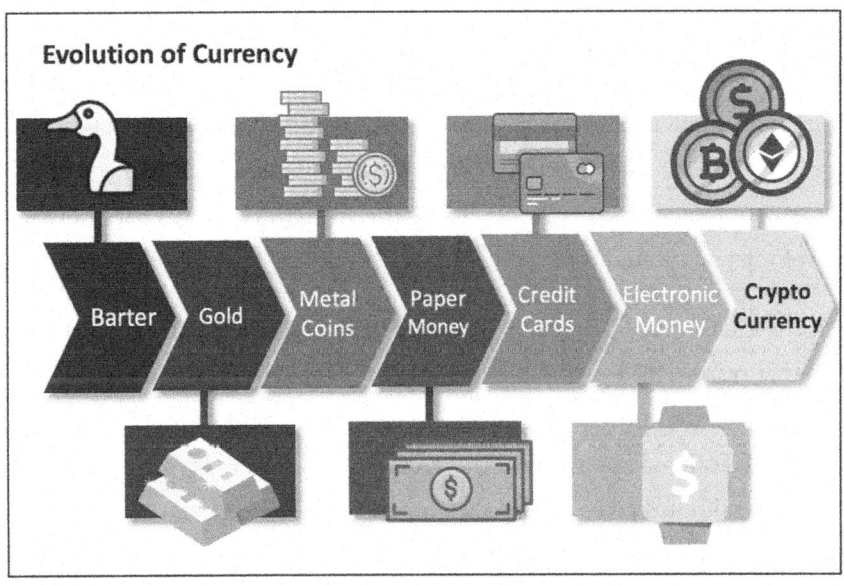

Chapter 4 – Figure 1 – The evolution of currency from duck bartering to crypto

However, for something to represent value, people must **trust** that it is indeed valuable and will stay valuable long enough for them to redeem that value in the future. For all but the last hundred years, we've always trusted in something to represent money (e.g., gold). Over this period, we've changed our trust model from trusting **something** to trusting in **someone**. People found it too risky and cumbersome to walk around the world carrying ducks or bars of gold. So, paper currency was invented.

Paper money works the following way. A bank or government offers to take possession of your bar of gold, let's say worth $100 and in return that bank would give you certificates, called bills, amounting to $100. Not only were these pieces of paper much easier to carry, but you could spend a dollar on a cup of coffee and not have to cut your gold bar into 100 pieces. And if you wanted your gold back, you simply took $100 in bills back to the bank to redeem them for the actual form of money, in this case, that gold bar. Hence, paper began its use as money as an instrument of practicality.

As time progressed the bond between the paper bill and the gold it stands for was broken. The macroeconomics that led to this change are complex but suffice it to say that governments worked out a deal with its people, where the government itself would be liable for the value of that paper money. This process works because of **trust**. Even though there is no actual commodity backing paper money, people trust the government and that's how fiat currency was created.

'Fiat' is a Latin word that means 'by decree,' meaning the dollar or euro or any other currency, for that matter, has value because the government declares it to. It's what's known as legal tender—coins or banknotes that must be accepted if offered as payment. Therefore, the value of today's money comes from a legal status given to it by the government. Hence, the trust model has changed from 'trusting something' to 'trusting someone,' in this case, the government.

Digital currency and the double spend problem

Once fiat money was in place, the move to digital money was pretty simple. We already have a central authority that issues money, so why not make money digital and let that authority keep track of who owns what. Today we mainly use credit cards, wire transfers, PayPal, Venmo, and other forms of digital money. The amount of physical money in the world is almost negligible and it's getting smaller with each year that passes. What then is slowing down the adoption of digital currency? Let's consider this example. If I have a digital file that represents a dollar, what's to stop me from copying it a million times and having $1,000,000? This is called the **double spend problem**.

Banks solve the double spend problem by creating a centralized solution. They keep a ledger on their computer, which keeps track of who owns what. Everyone has an account, and the bank ledger keeps a tally for each account. We all trust the bank and the bank trusts their computer.

There were many attempts to create alternative forms of digital currencies. However, none have been successful in solving the double spend problem without a central authority.

Rise of Cryptocurrency and Bitcoin

Fiat and digital currencies are seen to have several drawbacks that have given rise to cryptocurrency. The first is centralization. The users of today's currencies place their trust in a central authority, like governments or central banks, which control, issue and track the currency. Banks and governments gain your trust through reputation. Many of these institutions are hundreds of years old and have provable track records of ethical behavior. However, trust in centralized currency is sometimes compromised by corruption (e.g., Wells Fargo fake account scandal[23]), mismanagement (e.g., inflation due to over-printing of money[24]) and control (e.g., India abolishes 1000-rupee notes to fight corruption[25]).

Enter **Bitcoin** (BCT).

BITCOIN ECONOMICS

This was the state of currency until 2009. As we've discussed in the previous chapter, this changed with the publishing of the Satoshi Nakamoto paper. The document suggested a way of creating a system for a decentralized currency. This system claimed to create digital money that solves the double spend problem without the need for a central authority.

Bitcoin presents the industry's first transparent and decentralized ledger. At any point in time, we can see all the transactions and balances that are taking place on the ledger. In fact, it is simple for anyone to use the Bitcoin Explorer from blockchain.com to examine any and all Bitcoin transactions that have taken place since 2009. To embellish this point, here is the block that was just mined as I was writing this paragraph of this book.

Bitcoin Block 724476

This block was mined on February 22nd 2022, at 11:46 AM EST by Unknown.

The miner(s) of this block earned a total reward of 6.25 BTC ($233,882.56). The reward consisted of a base reward of 6.25 BTC ($233,882.56) with an additional 0.05819802 BTC ($2,177.84) reward paid as fees of the 1676 transactions, which were included in the block. The Block rewards, also known as the Coinbase reward, were sent to this address.

A total of 34,347.16 BTC ($1,285,312,087.57) were sent in the block with the average transaction being 20.50 BTC ($766,892.65).

Hash	00000000000000000027d610b2a7645d1d228250753 4a5504174e73b4b352ad
Timestamp	2022-02-22 11:46
Miner	Unknown
Number of Transactions	1,676
Difficulty	27,967,152,532,434.23
Merkle root	a424480f8649c19bda2864aeabee22a54b4d8bab4d18d90 9fb4dab4807dcc8f1
Nonce	3,648,339,987

Block Reward   6.25000000 BTC
Fee Reward   0.05819802 BTC

As this example shows, the transactions on the Bitcoin ledger are visible, however, the identities of those transacting are masked, preventing you from figuring out who owns these balances and who is behind each transaction. As discussed in the previous chapter, Bitcoin is pseudo-anonymous. Everything is open, transparent, and trackable, but you still can't tell who's sending.

There is an elegant supply-and-demand economics in play with Bitcoin. Satoshi put a hard cap or maximum coin-limit of 21 million on the Bitcoin supply, regulating it through an algorithm in its source code. As of today, over 90% of that fixed supply of coins have been issued. The limited supply makes it a scarce commodity and can help increase its price in the future.

A new Bitcoin is added to the supply at a fixed rate of one block every 10 minutes. However, the algorithm is such that the new bitcoins in each block are reduced by half every four years.

Around 19 million bitcoins have been mined, leaving only 2 million to be mined in the future. Experts predict that the remaining bitcoins will be mined by 2140.

Bitcoin also has a process that is used to introduce new coins into the existing circulating supply. Miners are integral in creating Bitcoin tokens; they solve cryptographic puzzles and validate a block of transactions in the network. For playing their part in the network, the miners get block rewards that include the newly-minted bitcoin and the cumulative transaction fees paid in a block. The only hurdle for miners is that the block reward also roughly halves every four years. In 2012, it was halved to 25 bitcoins, and it went down to 12.5 in 2016. From May 2020 through 2024, miners earn 6.25 bitcoin for every new block.

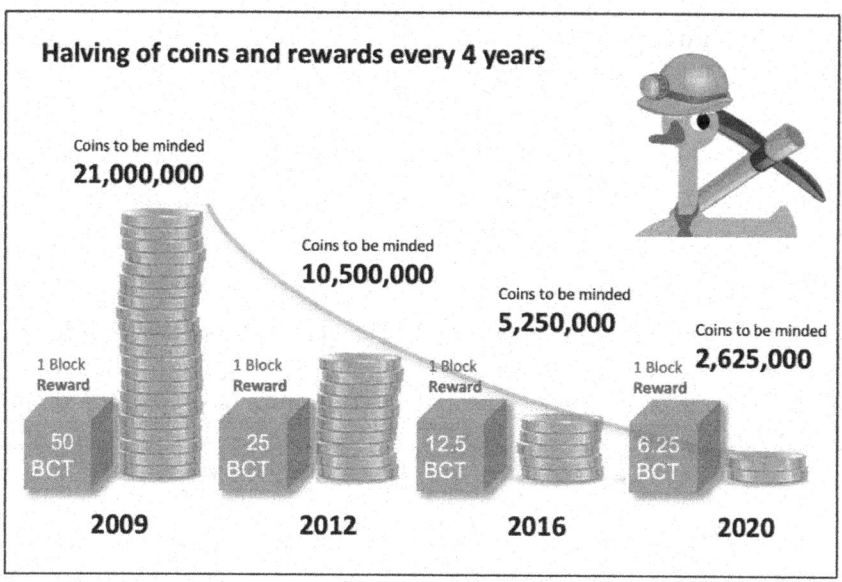

Chapter 4 – Figure 2 – Bitcoin mining reward and new mineable coins halve every four years

To summarize, the purpose of the Bitcoin network is to create a secure economy around the Bitcoin currency (i.e., bitcoins – By the way, as a concept Bitcoin is capitalized, however the currency unit, #bitcoin, is lowercase. I'm trying my best to keep it straight throughout this book, but pardon if I miss one or two) by enabling parties to transact by sending and receiving coins, while incentivizing miners to get rewarded with coins for doing work to maintain the integrity of the currency system. Therefore, to better understand how Bitcoin works, we need to understand these two primary personas. The next section takes a brief look at how parties **transact**. The section that follows with go deeper into **mining**.

TRANSACTIONS

A Bitcoin transaction is a transfer of bitcoin from one user to another. A transaction can be created in a crypto wallet, residing in the user's computer, smartphone, or tablet or in a cryptocurrency exchange (e.g., Coinbase.com). The transaction is published on the Bitcoin network where it is validated and added to the blockchain by Bitcoin miners (I promise we'll get to mining in a minute).

In very simple terms, a transaction is when party A gives a designated amount of bitcoin they own to party B.

Bitcoin makes use of public-key cryptography to ensure the integrity of transactions created on the network. In order to transfer bitcoin, each party has pairs of public keys and private keys that control pieces of bitcoin they own. We are not going to go very deep into public key infrastructure (PKI)[26] here but suffice to say, anyone who possesses your private keys has access to your bitcoin.

A **Bitcoin wallet address** is a hashed version of your public key. The public key is used to ensure you are the owner of an address that can receive funds.

To better illustrate how value is transferred in the Bitcoin network, we will walk through a sample transaction, where Donald duck sends .05 bitcoin to Howard duck.

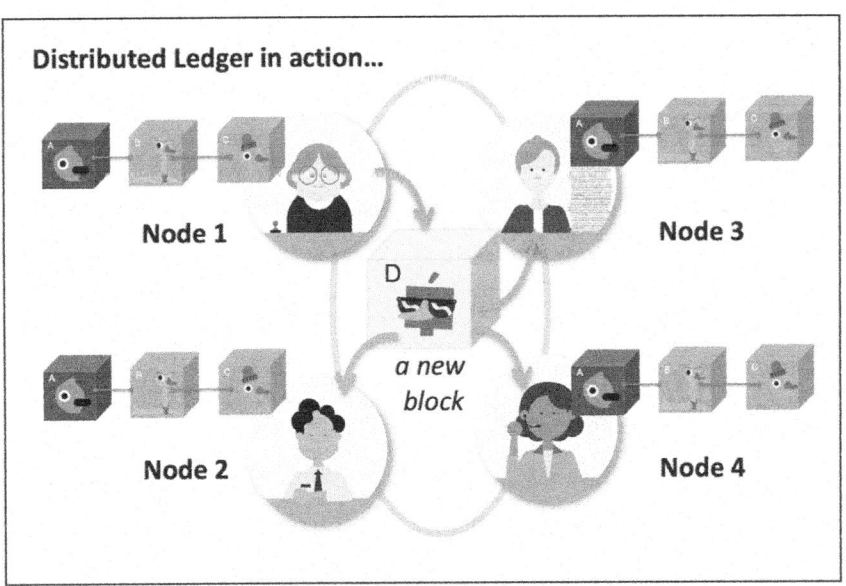

Chapter 4 – Figure 3 – Bitcoin transaction flow; Donald sends Howard 0.5 BTC

At a high level, a transaction has three main parts:

- **Inputs:** The Bitcoin address that contains the bitcoin Donald wants to send. To be more accurate, it is the address from which Donald had previously received bitcoin and is now wanting to spend.

- **Outputs.** Howard's public key or Bitcoin address.

- **Amounts.** The amount of bitcoin Donald wants to send.

Bitcoin does not have accounts and balances. Instead, pieces of bitcoin of arbitrary size are all associated with an address, which is controlled by the owner of that bitcoin. These pieces of bitcoin are called **Unspent Transaction Outputs** (UTXOs). Unlike a bank account balance, which is a single amount, a user's Bitcoin balance is like having cash stashed in different places throughout the house. Perhaps you have some cash in your wallet, some in a drawer, and some in your safe or under your mattress. The sum of all the money in your house is the spendable balance. In the case of Bitcoin, the sum of all UTXOs for a given user is that person's coin balance.

In the example above, notice that Donald is due 0.1 BTC in "change" from the transaction. However, he only receives 0.09 BTC because he owes the miner a transaction fee of 0.01 BTC.

To better understand how Bitcoin works, you need to understand how mining works. The next section shines a light on the process of mining and concludes with a short coding example, where we add mining to our Duckchain.

THE BITCOIN BLOCK

This section revisits the anatomy of a block. We introduced a simple definition of a block in the previous chapter. The Bitcoin application has a more elaborate block definition as seen in the sample **Block 724476** illustrated above. We will need to understand these properties to gain an appreciation of how mining works.

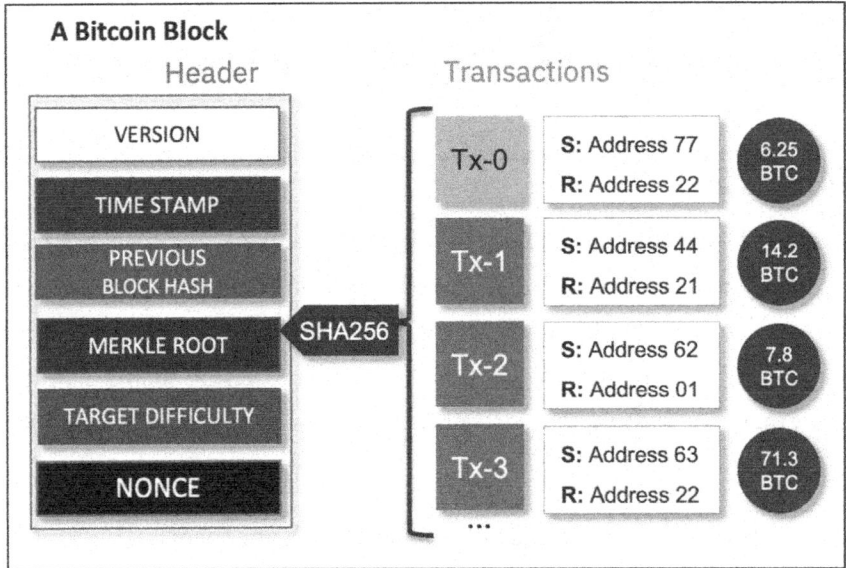

Chapter 4 – Figure 4 – The anatomy of a Bitcoin block

As the figure above shows, there are basically two facets to a Bitcoin block; the **block header** and the **transactions** included in that block. Each of these components is vital to creating an accurate and reliable header.

- A **version number** to track software/protocol upgrades.

- A **time stamp** is the approximate creation time of this block.

- The **previous block hash** links to the hash of the previous (parent) block in the chain

- The **Merkle root** is made up of all the hashed transaction hashes within the block's transaction list.

- The **difficulty target** is used to adjust how hard it is for the miners working to solve the block.

- The **nonce** is the value that miners can alter to create different permutations and generate a correct hash in the sequence.

These fields are not as complicated as they might sound and can be better appreciated when they are brought to light during the mining process.

MINING

A miner is represented by a node in the Bitcoin network that collects transactions and works to organize them into blocks. Mining can be viewed as a form of gamification. Miner nodes compete to create new blocks by collecting and aggregating new transaction data. Upon receiving such data, each node independently verifies each and every transaction against a long list of criteria. The list includes:

- tracking the source of the digital money being spent,

- checking for double spending of the same money,

- checking if the total transaction volume is within the allowed range of zero to 21 million bitcoins.

Bitcoin mining is one of the key elements that allows cryptocurrencies to work as a peer-to-peer, decentralized network without the need of a third-party central authority.[27]

Mining involves performing proof of work calculations and results in generating new coins as a reward for the miner, who successfully did this proof of work first. For each new block, proof of work requires a significant number of calculations aimed at solving cryptographic hash puzzles, with the lucky miner that ultimately mined that block receiving a reward in the form of bitcoin.

Specifically, the process of mining involves creating, transmitting, receiving, verifying, and packaging transactions.

Let me pause here and emphasize that Bitcoin is all about processing transactions. Specifically, the transfer of coins from user to user (i.e., wallet to wallet). Unlike our Duckchain, where the data field only contained a single entry (i.e., a name of a duck), the data field of the Bitcoin blockchain consists of a collection of transactions that can number in the one to two thousand range (which was the average in February 2022). Hence, the Bitcoin network not only verifies blocks, but also takes on the task of ensuring that all the transactions in the network are also valid. As you will see, this is done with even more hashing.

Transaction pooling

Sending a transaction on Bitcoin requires a small fee. Unlike most fees in life, you get to choose how much you want to pay—even $0.00. Here's the catch: supply and demand determine whether or not your transaction gets included in a block and executed.

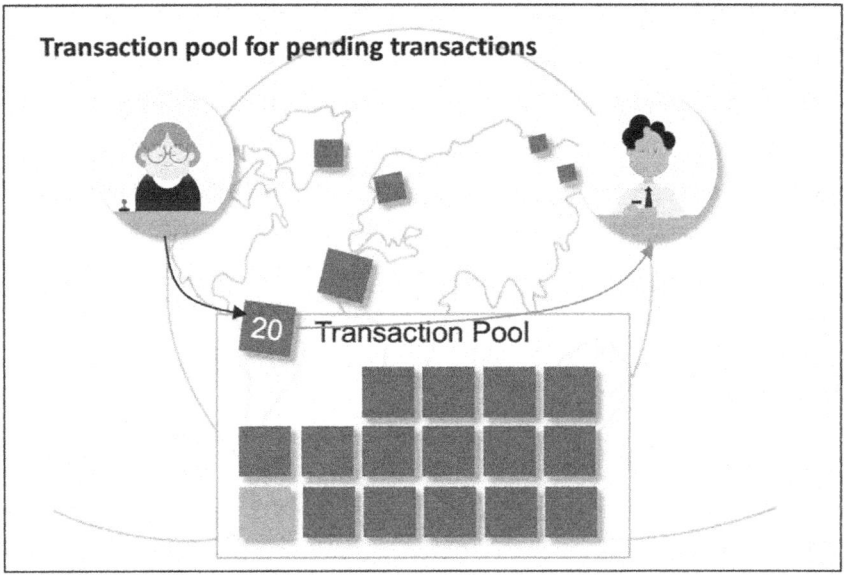

Chapter 4 – Figure 5 – New transaction sent to transaction pool before committed to block

For example, let's say you want to send $20 to a friend. Along with your $20, you decide to include $0.01 as a transaction fee. When you confirm this transaction, it is not instantly executed, but instead becomes a **pending transaction** and is added to the **transaction pool**—the group of all transactions that have been submitted, but not yet added to a block.[28]

Miners often choose the transactions with the highest fees to include in the next block. Of course, if there's no congestion on the network; miners will include all transactions that have been relayed to them.

Coinbase transaction

The miners that are successful in mining a block are in line for not one, but two rewards! These are referred to as the **block** and **fee** rewards.

The fee reward is the sum of the transaction fees of the transactions in the pool, described above.

The block reward (a.k.a., the mining reward) is the reward that is cut in half every 4 years (currently at 6.25 BTC). To ensure the miner gets this reward, the first block in the transaction pool is a transaction where "Satoshi" is the sender, the miner is the receiver, and the amount is set to the mining reward. This transaction is referred to as the **coinbase** transaction. (You can now appreciate where the company Coinbase.com got their name). It's a transaction where coins get created (from the pool of 21 million allocated by Satoshi) and in most cases, is the first transaction in a new block.

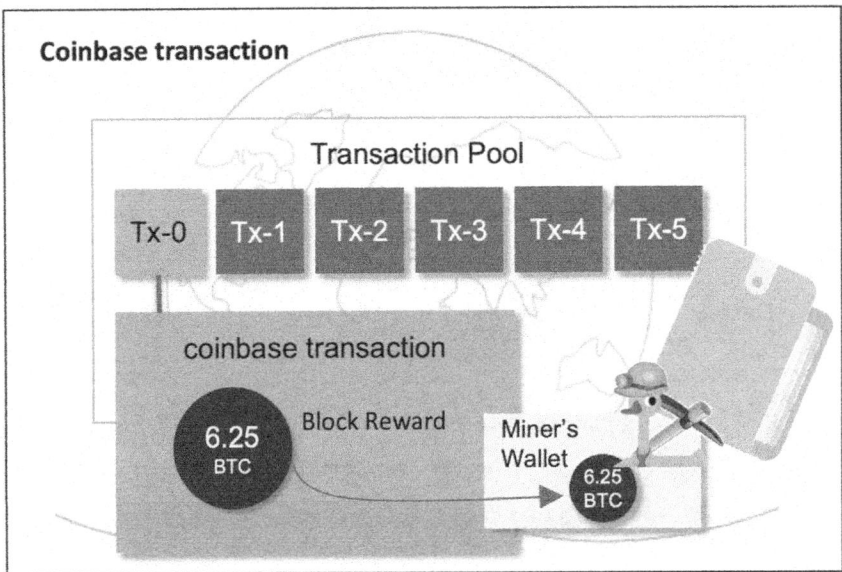

Chapter 4 – Figure 6 – Coinbase transaction ensures a miner gets paid the mining reward

As miners compete with each other to be the first one to come up with a new valid block, they need to make sure the transactions in their transaction pool have not already been included in previous blocks. This sounds like a job for a **Merkle tree** because it allows miners to verify a specific transaction without downloading the whole Bitcoin blockchain (over 324 gigabytes as of February 2022).

Merkle trees

The main advantage of using a Merkle tree in Bitcoin (as well as in most transaction processing blockchains) is that a set of transactions can be verified without the need to have access to the full data set of transactions.

Therefore, Merkle trees are an essential part of a blockchain for data integrity. Each block in the blockchain can contain multiple transactions, which each contain data.

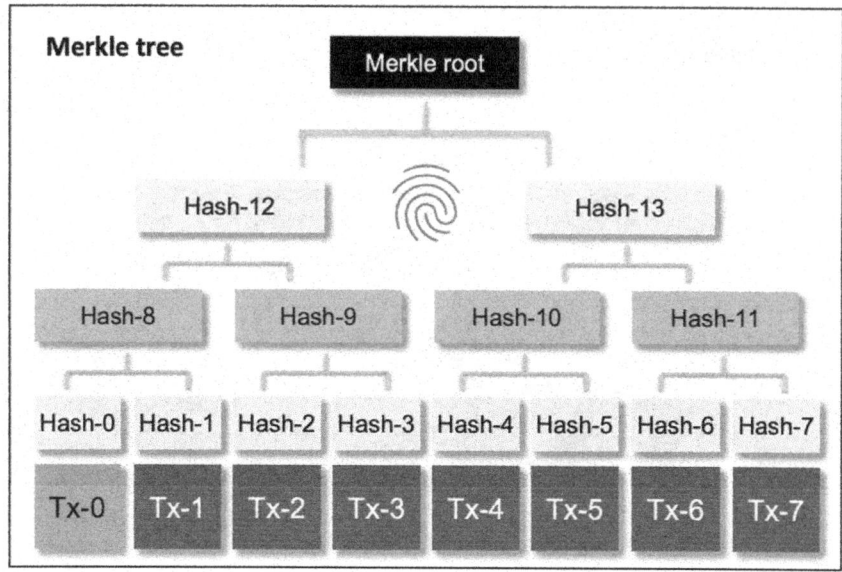

Chapter 4 – Figure 7 – Merkle root is the great grand-parent hash of candidate transactions

In this example, there are eight transactions. Each transaction's data is passed through a hash function, generating seven unique hashes, one for each transaction (i.e., hash 0-7). The eight hashes are paired to create "parent hashes" (i.e., hashes 8-11). These hashes are again hashed to form the "grand-parent hashes" (i.e., Hash 12 & 13). Finally, these two hashes are hashed to form the root hash, called the **Merkle root**, which is placed into the block header.

This results in a single root hash forming a complete Merkel tree. The Merkle tree allows for the detection of any changes within the transactions of a block by simply running through this process for each transaction and comparing the results to the original Merkle root.

Cryptographic puzzle

The miner is not done yet. It's time to play a game.

To solve the puzzle and win the prize, all a miner needs to do is to generate a block header with a hash value that starts with one or more zeros. The difficulty target is what dictates the required number of zeros. The more the number of zeros, the more difficult the puzzle is to solve.

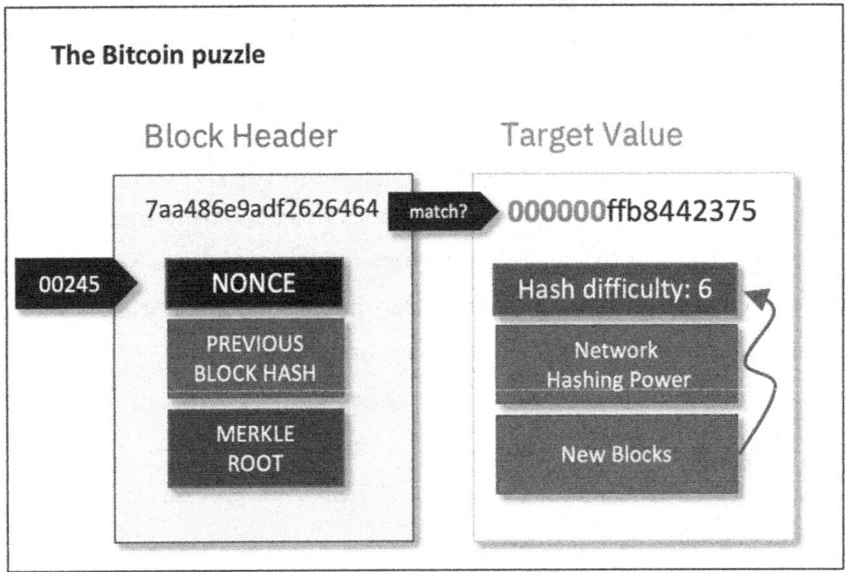

Chapter 4 – Figure 8 – Continuous incrementing nonce and
re-hashing match leading zeros needed for golden hash

If the miner used a (i.e., our favorite SHA-256) hashing function against the same header, without changing anything, they would get the same hash result, over and over again. This would be pointless. But this is where the nonce comes in. After each unsuccessful try to get a hash result with the right number of leading zeros, the miner increments (adds one to) the nonce and rehashes. This causes the resulting hash to be new and unique. Now, whether it has the right number of zeros is another story. It can take billions of iterations to generate the **golden hash value**.

This is basically a brute force approach of trying over and over again and is precisely the element of "work" that's attached to the term 'proof of work'. When a valid hash is found, the founder node will broadcast a block to the network. All the other nodes will check if the hash is valid and add the block into their copy of the blockchain ledger and move on to mining the next block.

A Tie

Sometimes it happens that two miners (e.g., miner A & miner B) broadcast a valid block at the same time and the network ends up with two competing blocks. This causes a temporary divergence in the network as miners start to mine the next block based on the block they received first (e.g., some mining the next block to follow A, others block B). The competition between these blocks will continue until the next block is mined (e.g., block C) based on either one of the competing blocks, the block that gets abandoned is called an orphan block, or a stale block. The miners of this block will switch back to mining the chain of the winner block, while the block reward is granted to the miner who discovers the valid hash first.

Mining Pools

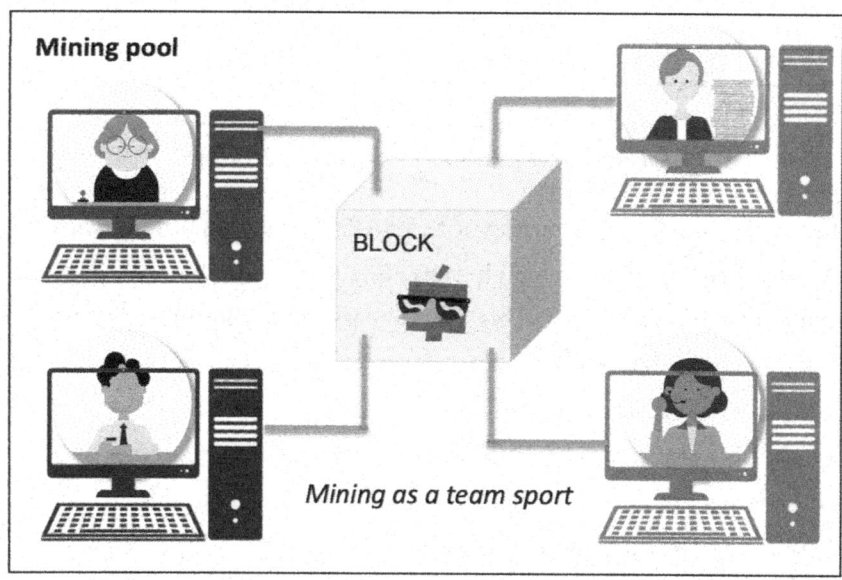

Chapter 4 – Figure 9 – Mining Pools – Mining as a team sport

The probability of finding the hash is equal to the proportion of the total mining power on the network. Miners with a small percentage of the mining power stand a very small chance of discovering the next block on their own.

Mining pools are created to solve this problem by pooling the computer resources of multiple miners over a network. Their processing power is combined to iterate through the crypto puzzle to find the golden hash of a new block. If found, they split the reward equally among everyone in the pool according to the amount of work they contribute to the probability of finding a block.

MINING CODE EXAMPLE

Once you understand the concepts of how the Bitcoin network operates, you start to appreciate some of the simple elegance behind the technology. However, I know it took me a couple of "go arounds" and what really help me is when I looked at a few code examples. This section will do just that. We will build on the Duckchain example from the prior chapter and add in some of the mining concepts that we've just reviewed.

Like before, this coding example takes a very simple path to illustrate the concept of mining. Therefore, we will skip the mining rewards aspect and look at the most basic properties of mining. The following code snippet illustrates a few additions required for mining.

```
 3    // This class defines a Block object destined for a Blockchain
 4    class Block {
 5
 6        // method that creates a new Block
 7        constructor(index, data, previousHash) {
 8            this.nonce = 0;                      // used to add variety to hash calc
 9            ...
10        }
..        ...
17        // method that calculates the hash of a block
18        // by applying a SHA256 hash function across block properties
19        calculateHash() {
20            let newHash = SHA256(this.index +    // include the index
21                JSON.stringify(this.data)  +     // all the data
22                this.nonce                 +     // include the nonce too
23                this.previousHash);              // and the previous hash
24            return(newHash.toString());          // returns a 64-character hash
25        }
26
27        // method to mine a new block by repeatedly calculating a new hash
28        // until the hash string starts with several "000" that match difficulty
29        // if no match found, increment the nonce which changes block content
30        // that ensures new hash calculation will differ from previous
31        mineBlock(difficulty) {
32            while (this.hash.substring(0,difficulty) !==
33                                    Array(difficulty + 1).join("0")){
34                this.nonce++;
35                this.hash = this.calculateHash();
36            }
37        }
```

The first addition, seen above on code-line 8, is the inclusion of a **nonce** property in the Block class. The example uses the nonce as a counter that is incremented by one during each iteration though the mining process. This simple technique ensures that each new block is different from the previous, giving each block a unique "fingerprint."

The *calculateHash* method needs to add the nonce to the DNA (so to speak) of the block. That is done on line 22. By the way, when I first coded this example, I forgot to add this line and the program just looped forever and my laptop got really hot because the fan was working overtime. Again, the nonce is important because it adds variability, as you will see in the next method discussed.

The Block class now also includes a *mineBlock* method, which performs the proof of work algorithm (a.k.a., the puzzle). As you can see, it's pretty simple. The method accepts a difficulty target as input. This value dictates what a new hash string must look like to win the puzzle matching game. For example, if the difficulty target is 1, the matching 64-character hash must start with one "0" to win. A target of 2 requires two "00"—and so on. This is what the code is doing on code-lines 32 and 33. If a match is not found, the loop keeps running. Line 34 increments the nonce by one. This is really important because if we don't change the block in question, it will keep generating the same hash value. So, with the new nonce, we recalculate the hash of the block (on code-line 35 above). Remember that if we change the Block just a little (like we just did), the resulting hash is random and potentially is now radically different from the previous value. From here, we loop back and check if the new hash has the zeros needed to win… if yes, we are done… if not, keep going.

Now let's look at the modification to the Blockchain class needed for mining.

```
39   // This class defines a blockchain object
40   class Blockchain {
41
42       // method that creates a new blockchain
43       constructor() {
44           this.difficulty = 1;
45           …
46       }
47       …
48       // adds a new block on to the chain
49       addBlock(newBlock) {
50           // fetch the hash of last block
51           newBlock.previousHash = this.getLatestBlock().hash;
52           // calc has of this new block
53           newBlock.mineBlock(this.difficulty);
54           // add the new block to chain
55           this.chain.push(newBlock);
56       }
57
```

The updated Blockchain class has a new property, which relates to the target **difficulty** that was just discussed in the code above. We will see this property in action soon.

There is one important adjustment needed to the ***addBlock*** method. In the previous version we simply calculated a new hash of the block passed in as input. In this updated version we call the new mineBlock method and pass in the difficulty target… and wait for proof of work to complete with a winning Block that fits the criteria of "the puzzle" (i.e., has the right number of leading zeros).

Okay, now let's create some code to test out this new mining operation.

```
87  // create and test the Duckchain
88  let duckchain = new Blockchain(); // Tada! duckchain is live on my laptop
89
90  duckchain.difficulty = 3;
91  duckchain.addBlock(new Block(1, "donald"));  // add the donald block
92
93  duckchain.difficulty = 6;
94  duckchain.addBlock(new Block(2, "daffy" ));  // add the daffy block
95
96  // print out the full duckchain
97  console.log(JSON.stringify(duckchain, null, 4));
98  // print out if chain is valid or not
99  console.log('Is the Duckchain valid? ' + duckchain.isChainValid());
```

Similar to the previous chapter, the test starts by instantiating a new Blockchain object called **duckchain**, followed by code that adds new blocks. Before new Blocks are created and added, the difficulty is set. First, it's set to a difficulty target of 3 and then a target is doubled to 6. Once the two new blocks are mined and added, we display the entire contents of the duckchain object to the console and finish by checking the validity of the chain.

The results are quite interesting.

```
{
    "chain": [
        {
            "index": 0,
            "data": "Genesis Block Quack Quack!",
            "nonce": 0,
            "hash": "7aa486e9adf2626464fd8e7e3abae661
                    4ad0c8d587dd5b95fd2809755c439718",
            "previousHash": "0"
        },
        {
            "index": 1,
            "data": "donald",
            "nonce": 5166,
            "hash": "000ad18b8f20f6f0f38d055514288689
                    43ebd3c2e35baea563b5fe5287423233",
            "previousHash":
                    "7aa486e9adf2626464fd8e7e3abae661
                    4ad0c8d587dd5b95fd2809755c439718"
        },
        {
            "index": 2,
            "data": "daffy",
            "nonce": 30141575,
            "hash": "000000ffb8442375617ee41ce5971983
                    b0ba07e722ee6e03e3fe446618812ac5",
            "previousHash":
                    "000ad18b8f20f6f0f38d055514288689
                    43ebd3c2e35baea563b5fe5287423233"
        }
    ],
}
Is the Duckchain valid? true
```

As you can see above, Block 1 (Donald) was mined with a hash difficulty of 3, so it has 3 leading zeros. Block 2 (Daffy) was a hash that starts with 6 zeros because we set the difficulty target accordingly.

I want to draw your attention to the nonce values associated with these blocks. The nonce property of the Donald block has a value of 5166. Which means mining this block required 5166 iterations before hitting the lucky number that started with 3 zeros. Executing this proof of work was all but instantaneous. On the other hand, it took a few minutes to execute the proof of work for the Daffy block. With a difficulty target of 6, it took 30,141,575 iterations before hitting the golden hash with 6 leading zeros. In fact, my laptop fan was really humming when running this exercise and got me thinking about the energy efficiency (or lack thereof) of this algorithm.

ALTCOINS AND BEYOND

An Altcoin is an alternative digital currency to bitcoin. As of today, over 5000 of these "alternative" currencies have been created worldwide. Most of the Altcoins are based on Bitcoin, and their basic functions are essentially the same. Some of the most well-known Altcoins (based on market cap) are Ethereum, Ripple, Tether, Bitcoin Cash, Bitcoin SV, and Litecoin.[29]

Litecoin (LTC) is a good example of a "derivative" of Bitcoin. It was created in 2011 from a copy of Bitcoin's source code with modifications aimed at addressing some of Bitcoin's shortcomings. This makes Litecoin an entirely new blockchain network without a shared genesis block. Litecoin is technically very similar to Bitcoin but has a few crucial differences. Specifically, Litecoin features:

- A unique Litecoin mining hashing function

- A faster block generation time

- Increased maximum coin supply[30]

As you can see, there is no one single blockchain to rule them all. While Bitcoin should always be appreciated as the blockchain that put crypto on the map, crypto has and will continue to evolve in interesting ways.

QUIZ TIME

Bitcoin is the application that made blockchain famous. This chapter touched on many of the key facets of how Bitcoin utilized blockchain to create currency with a working economy. Picking quiz questions was not easy given all the ground we covered. But with the chapter as input, here is the "unspent quiz question output" (UQQO).

1 – Bitcoin is a digital payment system that uses decentralized management to avoid?

    a. Clerical errors that often occur when processing paper documents

    b. Double spend

    c. Peer-to-peer payments used for check splitting

    d. Large transaction processing fees

2 – What does it mean when we say that Bitcoin is pseudo-anonymous?

    a. Administrators are anonymous

    b. Everything is open, transparent, and trackable, and you can tell exactly who is transacting

    c. Everything is open, transparent, and trackable, but you still can't tell who's transacting

    d. None of the above

3 – Miners independently verify the integrity of transaction by?

    a. Tracking the source of the digital money being spent

    b. Checking for double spending of the same money

    c. Checking if the total transaction volume is within the allowed range of zero to 21 million bitcoins

    d. All of the above

4 - The proof of work is done by matching a golden hash value with the correct number of leading zeros by hashing and

re-hashing the block's header and incrementing (often billions of tries) the block's nonce after each unsuccessful iteration?

a. True

b. False

5 - Think Blockchain Labs – Chapter 4 Assignment

Go to the Think Blockchain Labs Github repository.

See the README.md file and search for the Chapter 4 assignment details, but the gist of the assignment is to:

a. Run the simple Duckchain sample, that now includes mining

b. Enhance the Duckchain sample by modifying it to include a start and stop time log message and calculate elapsed time of mining.

c. Re-run sample several times, changing the difficulty field and note how the time increases for each difficulty level added.

d. As a bonus, add a method that calculates an approximation of energy used to mine a new Duckchain block.

UP NEXT

To revisit the method behind the madness of this book, in Chapter 3 we abstractly defined how blockchain works. In Chapters 4, 5 and 6, we set out to study three pioneers of blockchain technology that have paved the way by introducing a game-changing concept that stretches the state-of-the-art of distributed ledger technology. This chapter was all about how cryptocurrency was invented with a clever combination of consensus technology atop a novel economic model of rewards, which established Bitcoin as the first pioneer of blockchain. In fact, here is a photo from my blockchain camera roll of a cryptocurrency server farm mining bitcoin. If you look close enough, you can see the GPU graphic-card rigs used for cryptocurrency mining.

Chapter 4 – Figure 10 – Cryptocurrency server farm

I can't resist sharing one more photo. This collage is of a series of photos that went along with blockchain articles featuring yours truly in the online magazine cointelegraph.com in 2016. *Cointelegraph* covers the latest news in crypto and is known for the hilarious caricature-style of headline art that accompany the articles.

Chapter 4 – Figure 11 – CoinTelegraph 2016 articles featuring Jerry with caricatures

- - -

Our next chapter looks at pioneer #2, covering how the introduction of smart contracts enables blockchain technology to become multi-purposed beyond cryptocurrency to digitize value in most every form.

Chapter 5 –
TOKENS, ETHEREUM, AND SMART CONTRACTS

Reports project that the market cap of the Non-Fungible-Token (NFT) industry will approach US$80 billion by 2025. Ethereum's smart contracts have ushered in the new, cool kids on the block in crypto.

COVERED IN THIS CHAPTER

- A personal history of tokens

- Taxonomy of digital assets

- Ethereum as a world super-computer

- Smart Contracts and Ethereum Tokens

- Coding your own Token and NFT

EXPANDING THE AMBIT OF BLOCKCHAIN

In the previous chapter, we discussed how money can be transformed with the help of blockchain. This begs the question; what other functions of society could benefit? Well, consider how the following applications would also benefit from blockchain's properties of transparency and immutability:

Voting – counting and validating votes

Real estate – clear provenance of ownership and transferring of records

Social networks – self-managed digital rights of data

This chapter expands on the usage of blockchain as seen in Bitcoin to other types of digital assets. Like the previous topic, we will tackle this subject by looking at three related topics:

Tokens – going beyond currency to any digital asset

Ethereum – as the network that pioneered the creation of a world computer for digital assets

Smart contracts – as the defining technology that enables our "token economy."

We start with the evolution of tokens and walk through my memory lane of the history of tokens leading to today's digital tokens. Here goes…

MY HISTORY WITH TOKENS

When I was a kid, back in the 1970s, arcades were a popular place to hang out. The game machines at the arcade typically operated on tokens. These tokens were brass coins that you would insert into the video games in order to play them. Typically, they would be for sale at four for a dollar, acting much like quarters, but if you inserted a $5 bill in the change machine, you could get twenty-five tokens (a $6.25 value). The rest was "Pac-Man" history.

Of course, you didn't have to use all your tokens in one day. You could take them home and the owner would never care. This is because, outside the arcade, the tokens had no real value. The owner used tokens as a means to create an economy within the arcade ecosystem. For me, the value of a token was a means to have fun for an hour or so, with an upside potential to win prizes. For the owner, the value of my purchasing $5 of tokens was to offset the fixed cost of owning and maintaining arcade machines, and to fund the prizes, which often cost far less than $5 to buy (think .50 cents) and hopefully make a profit.

Growing up in New York City, I had experience with another interesting type of token. While arcade tokens had little value outside of the arcade, subway tokens were different. Like the arcade tokens, the subway tokens had utility; they provided you access to the subway. One token, one ride. Unlike the arcade, however, subway tokens were necessary in NYC. Everyone needed tokens, and most people used them daily. As a result, it was always easy to find a buyer for tokens. In some places, you could barter with them to buy things (like arcade tokens). In addition to providing utility, the tokens were also acting as a form of currency.

As I got older, I discovered how to lose money in the "city of lost wages." Taking my history of tokens a step further, one also sees tokens at a casino. Here, the tokens are not used as a utility but to represent cash amounts. They become a currency that can be used within the casino. In the early days of Las Vegas, the casino tokens were freely used as currency throughout the city; the IRS has since outlawed the practice. Also, tokens at one casino are not accepted at a competing casino.[31]

In today's world, tokens have gone digital. Over time, many places that used to use tokens have switched to cards or mobile apps. For example, my favorite arcade today is Boxcar in downtown Raleigh, North Carolina. (What's not to like? It's a bar and arcade). Boxcar has games that operate using magstripe cards. The idea of the utility tokens remains the same, but the physical tokens are now replaced with data.

Since tokens can be represented by data, they can live online and be transmitted over the Internet. These computer-based tokens live on the web and provide a variety of purposes.

My (our) life experience with tokens is now starting to get very interesting. Blockchain technology is on the brink of opening the door for a new economy based on digital tokens. Hence, as a gateway into the future of this new economy I will wrap up this section by illustrating my favorite type of tokens, which are **loyalty tokens.** In fact, as we speak my (mobile app) wallet is literally littered with a collection of loyalty cards; airlines, hotels, coffee, clothing, groceries, and the list goes on and on. When you think about it, the digital points, levels, and status all have significant value to the "token" bearer. Heck, if I were to somehow lose these tokens, I'd be as devastated from the loss as I would feel for any cash or credit card loss. I had to drink a lot of coffee and spend many nights away at hotels to earn this value. I don't think any of you will debate this, however, you might also agree that while these digital assets have value, they are not money.

One important aspect is they are only valuable in the "ecosystem" within which they were created. You cannot easily exchange hotel points for airline points. And if there was, what would be the basis for the exchange? Does one airline point equal two hotel points? Is there a marketplace or digital wallet vendor that would facilitate such exchanges? And what about the technology underpinnings? As a technologist, I often think about the data structures, the contract logic and the on-wire protocols that would enable such exchanges.

The next section takes a closer look at these digital tokens and assets and the technology standards that are emerging around blockchain to progress towards this new economy.

DIGITAL ASSETS

In the mid-1990s, "digital assets" originated as a term encompassing items such as audio, video, images, and documentation. In the decades since, technological advancements have expanded the term's use to cover a broader range of items. Blockchain plays a critical role in the rapidly digitizing global economy.[32]

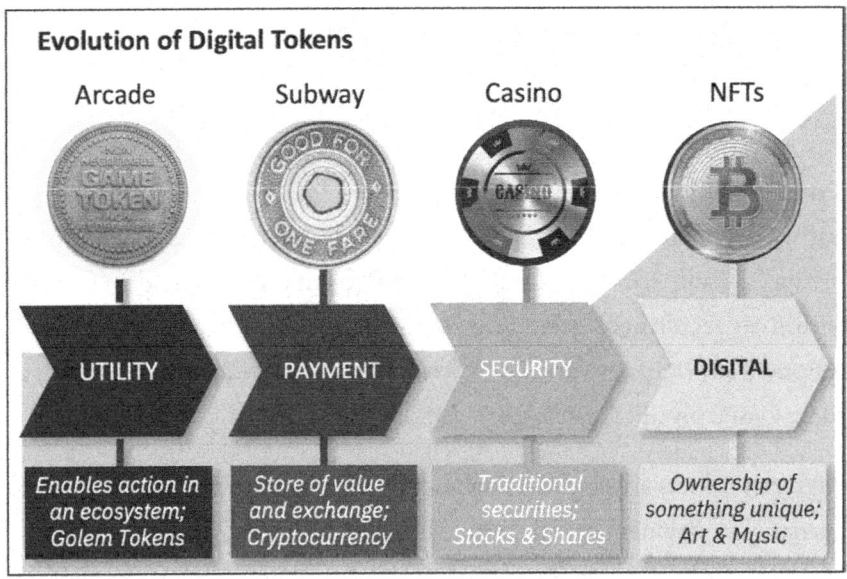

Chapter 5 – Figure 1 – Evolution of tokens to today's digital tokens

Modern digital assets are often expressed as cryptographic tokens, which represent a digital unit of value that lives on the blockchain. Familiarity with the concept of fungibility in economics might help one better understand how tokens represent value.

Fungible tokens are divisible and non-unique. For instance, fiat currencies like the dollar are fungible: A $10 bill in Raleigh has the same value as a $10 bill in Austin. A fungible token can also be a cryptocurrency like Bitcoin: 1 BTC is worth 1 BTC, no matter where it is issued.

Non-Fungible Tokens, on the other hand, are unique and non-divisible. They should be considered as a type of deed or title of ownership of a unique, non-replicable item. For example, a concert ticket is non-fungible because there cannot be another of the same kind due to its specific data (e.g., date, time, row, seat). A house, a boat or a car are non-fungible physical assets because they are one-of-a-kind.

There are four main token types as depicted in the previous figure above: **Utility, Payment, Security, and NFT tokens.** The following section outlines each token type.

Types of Digital Tokens

Utility tokens are similar in spirit to the arcade token described in the prior section; a utility token is a fungible crypto token that serves some use case within a specific ecosystem. These tokens allow users to perform some action on a certain network. Utility tokens give the holder access to a block-chain-based product or service.

Payment tokens are similar to subway tokens, which are both a utility and currency. A payment token shares these qualities and is best exemplified by today's cryptocurrencies. Their main purpose is to serve as a medium of exchange, store of value, and be a unit of accounting. Major cryptocurrencies discussed in the previous chapter, like Bitcoin and Litecoin are fungible payment tokens. Like fiat currencies, payment tokens gain or lose value based on the laws of supply and demand. Greater demand and lower supply increase value, while lower demand and greater supply decrease value.

Most cryptocurrencies have a finite supply. As discussed, only 21 million bitcoins can ever be mined. This means that, as more people start paying for goods and services with cryptocurrencies and the supply of new coins dwindles, their value should rise sharply, at least in theory.

Security tokens are traditional assets like stocks and shares that have been converted into a digital token on a blockchain. Like traditional securities, security tokens give the holder ownership rights. For this reason, a growing number of regulators are controlling how they're to be issued and traded. Even our loyalty points, as in the example mentioned above, would be best represented as security tokens. Consider how an airline might mint 100 million security tokens coded to represent miles (in a one-to-one ratio) and just hold them. As passengers complete their flights, the airline can automatically disperse the proper number of miles into each passenger's connected digital wallet.

Non-Fungible Tokens (NFT) are a digital representation of something unique and are considered the new cool kids on the block in crypto. Reports project that the market cap of the NFT industry will approach US$80 billion by 2025.[33] As stated above, NFTs represent digital ownership of a wide range of irreplicable intangible items. Each token represents a specific asset, so there's no standard value. This means you can't exchange one Non-Fungible Token for the other directly. That said, because of data that lives on a blockchain's distributed ledger, they can't be duplicated or altered, hence are ideal for proving ownership rights, identity, and authenticity.

Minting digital assets

In summary, cryptographic tokens are different from cryptocurrency. They have greater utility than coins, which are used solely as stores of value and currency. Even things that you might not think of as having a digital representation can be represented by digital tokens. That includes things like artwork and consumables.

Digital assets are minted on a blockchain and brought to life through smart contracts, which we will dive into later in this chapter. But before we visit smart contracts and look closer at digital asset examples, we will benefit from studying the blockchain network that pioneered the smart contract and is now the most popular place to mint new digital assets.

THE ETHEREUM WORLD COMPUTER

The Bitcoin application is written in what is known as a Turing Incomplete language.[34] By design, Bitcoin is simple and supports only the most basic instructions to process transactions. However, processing the creation and exchange of arbitrary digital assets would require a more complex system, with a more complete programming language. The challenge is how do you process a Turing Complete system that works at the same scale of Bitcoin, while preserving the transparency and immutability properties? Answer: **Ethereum.**

Ethereum is a decentralized blockchain platform that establishes a peer-to-peer network that securely executes and verifies application code, called **smart contracts**. Smart contracts allow participants to transact with each other without a trusted central authority.

Ethereum was first proposed in late 2013 and then brought to life in 2014 by Vitalik Buterin, who at the time was the co-founder of *Bitcoin* magazine. Ethereum is the do-it-yourself platform for decentralized programs. Also known as decentralized applications or **DApps**.

DApps can be programmed in a special programming language called **Solidity**. This (Turing Complete) language was specifically created for Ethereum and uses a syntax that resembles JavaScript. It's also worth mentioning that many of today's blockchain technologies support decentralized applications. In fact, Bitcoin also has support for smart contracts, although it's a lot more limited compared to Ethereum. But it's fair to say that Ethereum pioneered the DApp concept by allowing the creation and deployment of these decentralized programs that no single person controls, not even you, even though you wrote it. Hence, we will study Ethereum's support for DApps a little further in the sections to follow.

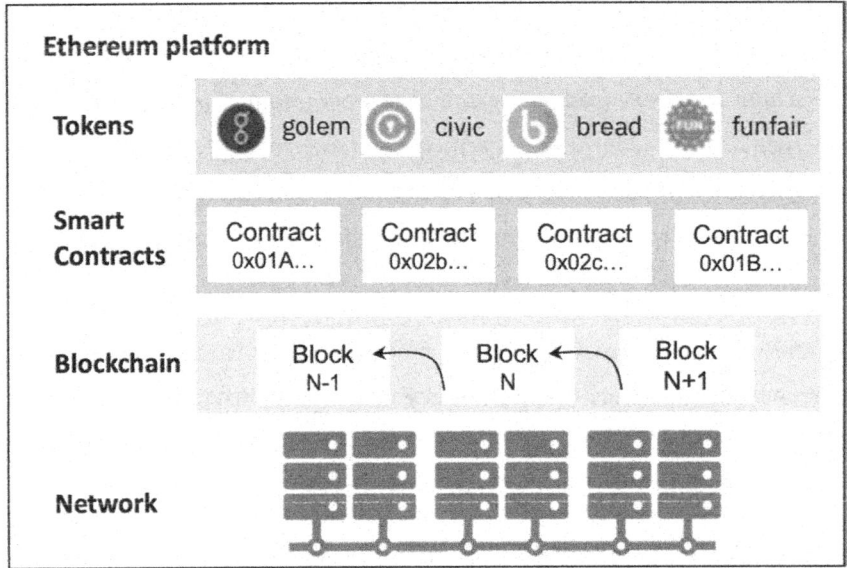

Chapter 5 – Figure 2 – Ethereum platform architecture

Like Bitcoin, the Ethereum platform has thousands of independent computers running it. Vitalik describes it as a "world computer" in his paper, "A Next-Generation Smart Contract and Decentralized Application Platform." Here is the abstract from his now famous paper:

Satoshi Nakamoto's development of Bitcoin in 2009 has often been hailed as a radical development in money and currency, being the first example of a digital asset which simultaneously has no backing or "intrinsic value" and no centralized issuer or controller. However, another, arguably more important, part of the Bitcoin experiment is the underlying blockchain technology as a tool of distributed consensus, and attention is rapidly starting to shift to this other aspect of Bitcoin. Commonly cited alternative applications of blockchain technology include using on-blockchain digital assets to represent custom currencies and financial instruments ("colored coins"), the ownership of an underlying physical device ("smart property"), non-fungible assets such as domain names ("Namecoin"), as well as more complex applications involving having digital assets being directly controlled by a piece of code implementing arbitrary rules ("smart contracts") or even blockchain-based "decentralized autonomous organizations" (DAOs). What Ethereum intends to provide is a blockchain with a built-in fully fledged Turing-complete programming language that can be used to create "contracts" that can be used to encode arbitrary state transition functions, allowing users to create any of the systems described above, as well as many others that we have not yet imagined, simply by writing up the logic in a few lines of code.[35]

Ethereum is the infrastructure for running DApps worldwide. It's not a currency, it's a platform. The currency used to incentivize the network is called **Ether**. As Vitalik implies, Ethereum is a network of computers that together combine into one powerful decentralized supercomputer.

Ethereum's coding language, Solidity, is used to write smart contracts that are the logic that runs DApps. But before we forge ahead into DApp development in Solidity (and I promise we will), it's important that we pause briefly and gain a more complete understanding of this notion of "smart contracts." We will come back to Ethereum in a few minutes.

SMART CONTRACTS

A smart contract isn't unlike its paper predecessor. It helps you exchange property, services, and currency. But unlike that hardly-enforceable paper stack just barely stapled together, this contract is a self-executing document.[36]

A smart contract is an autonomous (i.e., self-executing) contract, where the terms of the agreement between buyer and seller are directly written into a simple program (i.e., code) that resides on a blockchain's distributed ledger. Smart contracts permit trusted transactions and agreements to be carried out among disparate parties that render transactions traceable, transparent, and irreversible.

Smart contracts were first proposed in 1994 by **Nick Szabo**, an American computer scientist who invented a virtual currency called Bit Gold in 1998, ten years before the invention of Bitcoin. In fact, Szabo is often rumored to be the real Satoshi Nakamoto, the anonymous inventor of Bitcoin, which he has denied. (See the "I am Satoshi" Supplemental Chapter). Szabo defined smart contracts as computerized transaction protocols that execute terms of a contract. He wanted to extend the functionality of electronic transaction methods, such as point of sale, to the digital realm. Here is an excerpt from Nick's paper, "The Idea of Smart Contracts," where he likens smart contracts to vending machines.

> *A canonical real-life example, which we might consider to be the primitive ancestor of smart contracts, is the humble vending machine. Within a limited amount of potential loss (the amount in the till should be less than the cost of breaching the mechanism), the machine takes in coins, and via a simple mechanism, which makes a freshman computer science problem in design with finite automata, dispense change, and product according to the displayed price. The vending machine is a contract with bearer: anybody with coins can participate in an exchange with the vendor. The lockbox and other security mechanisms protect the stored coins and contents from attackers sufficiently to allow profitable deployment of vending machines in a wide variety of areas.*[37]

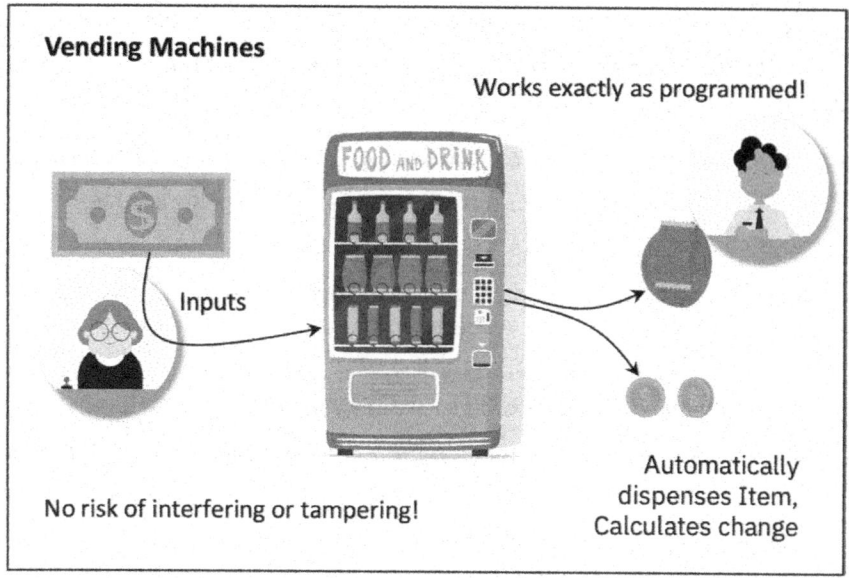

Chapter 5 – Figure 3 – Vending Machine as a Smart Contract metaphor

As the image above illustrates, the usage process of a vending machine is split into three distinct steps.

1. *Buyer selects item on the screen and agrees to the specified payment.*

2. *Buyer inputs (hard-earned) cash into the vending machine.*

3. *The machine recognizes the payment, confirms its validity, and drops the buyer a bag of peanut-butter and chocolate goodness in the bottom slot.*

As Nick Szabo infers, smart contracts are quite literally just repurposed, online vending machines. Each contract is designed to oversee a specific type of digital transaction. If the parties involved in the trade provide the necessary value required to fulfill the smart contract parameters, the transaction goes through, and both parties receive what they want. No middleman. No traditional transaction-bound costs. No risks.[38]

Now that we understand a little of the theory behind smart contracts, the next section will elaborate on how they work.

How smart contracts work

Later in this chapter, we will write a set of simple Ethereum smart contracts. To prepare for that, it will help to better define a smart contract and its relationship to blockchain.

As stated above, a smart contract is a digital contract, written as code and stored as data on a blockchain distributed ledger. Specifically, the smart contract has details and permissions written in code that require an exact sequence of events to take place to trigger the agreement of the terms mentioned in the smart contract. It can also include the time constraints that can introduce deadlines in the contract.

There is a powerful elegance gained by having the contract cryptographically hashed on a blockchain ledger, making it transparent, immutable, and distributed. Being immutable means that once a smart contract is created, it can never be changed again. Meaning no one can accidentally or maliciously tamper with the code of your contract. Being distributed means that the output of the contract is validated by peers on a blockchain distributed network. Hence, a single person cannot force the contract to "release the funds" because other validators on the network will spot this attempt and mark it as invalid. Like any data on a blockchain, tampering with smart contracts becomes almost impossible.

How smart contracts work

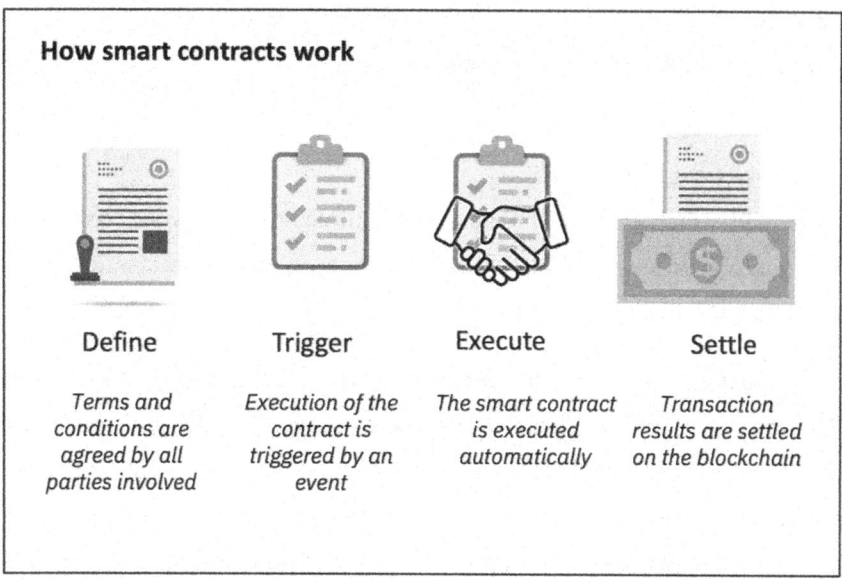

Define	Trigger	Execute	Settle
Terms and conditions are agreed by all parties involved	*Execution of the contract is triggered by an event*	*The smart contract is executed automatically*	*Transaction results are settled on the blockchain*

Chapter 5 – Figure 4 – Smart Contracts follow a process of Define, Trigger, Execute & Settle

Let's look at the steps involved in developing a smart contract.

Define – The process begins with a contract specification that outlines the abstract terms and conditions of the contract. This process is similar to the creation of a paper contract, but in this case, it involves collaboration between people with business skills and people with blockchain development skills.

Participants must agree on the rules governing those transactions, investigate possible exceptions, and create a framework for resolving conflicts to set terms. There can be as many criteria as needed within a smart contract to convince the parties that the work will be executed correctly.

Trigger – The smart contract's target behavior can be triggered in response to a variety of events or conditions. Conditions such as payment authorization or a shipment receipt are examples of simple events for which businesses use smart contracts. More complicated events, such as automatically delivering an insurance payout in the case of a person's death or a natural disaster, can be encoded using more complex logic.

Execute and Settle – The developers then create the logic and test it on a smart contract development environment to verify its execution. After the application is written, it needs to be reviewed. Often a security review panel is put in place to ensure robustness of a contract. This could be an internal professional or a business specializing in vetting smart contract security. Once validated, the smart contract is implemented on an existing blockchain.

Once implemented, the smart contract is set to listen for event updates from an "oracle," which is effectively a cryptographically protected streaming data source. The smart contract executes when it acquires the right mix of events through one or more oracles. We will discuss blockchain oracles in the Web3 Chapter of this book.

ARE SMART CONTRACTS REALLY SMART?

Going back to my previous example of the vending machine, in the non-digital world, if a customer puts their money in, and nothing comes out because the candy bar "got stuck," there is always recourse. You might try shaking the machine. (I do not recommend this option). You can take a photo of the machine and call the "1-800" number listed on the machine to reach a human and explain your situation. But there are no "1-800" numbers for smart contracts. They can't think or talk back. In short, by design, they lack "smarts."

A truly intelligent contract, on the other hand, would take into account other factors as well, such as extenuating circumstances and the spirit with which the contract was written. It would also be able to make exceptions if warranted. In other words, it would act like a really good referee. Instead, a smart contract in the context of Ethereum is not really intelligent at all. It's actually uncompromisingly letter strict.

It follows the rules down to the letter and can't take any secondary considerations or the spirit of the law into account like what commonly happens with real-world contracts. Once a smart contract is deployed on the Ethereum network, it cannot be edited or corrected. Even by its original author (remember, it's immutable). The only way to change this contract would be to convince the entire Ethereum network that a change should be made, and that is virtually impossible.

This introduces a potentially serious problem, especially when you consider that unlike Bitcoin, Ethereum supports the ability to create complex contracts. And complex contracts are very difficult to secure and debug. As with any contract, the more complicated it is, the harder it is to enforce as more room is left for interpretation or more clauses must be written to deal with contingencies. With smart contracts, security means handling with perfect accuracy every possible way in which a contract could be executed. Ethereum was launched with the idea that **code is law**. That is, a contract on Ethereum is the ultimate authority, and nobody can overrule the contract.

The Ethereum DAO

On June 17, 2016, the "code is law" ideal was put to the test when an Ethereum token, called DAO, was compromised and significant funds were lost. DAO stands for Decentralized Autonomous Organization. It was launched in April 2016 after a crowdfunding campaign via the DAO token sale. DAO became one of the largest crowdfunding campaigns in history, raising $150 million in Ethereum currency when Ether was trading at around $20. While this was an interesting idea, the code had vulnerabilities that were ultimately exploited, draining the DAO and its associated ether with it, to other unknown accounts.

Now you could say that the person who drained the DAO was a hacker, but some would argue that the hacker was taking advantage of the loopholes they found in the DAO's smart contract. And this isn't very different than a creative lawyer figuring out a loophole in the current law to affect a positive result for his client.

The Ethereum community reacted strongly and ultimately bended the rules of "code is law" and changed the Ethereum network in order to revert all the money that went into the DAO. The small minority that didn't agree with this decision kept to the original Ethereum blockchain before its protocol was altered. And that's how Ethereum Classic Network was born, which is actually the original Ethereum network.

ETHEREUM TOKENS...

Now that we've covered the basics of Ethereum, smart contracts and digital tokens, we are ready to continue on to the topic of Ethereum tokens. So far, we've talked about Ethereum as a world computer that powers the ability to both commit code to its blockchain as well as execute that code across the world-wide distributed network. Tokens require "gas" to run. We have yet to discuss in detail the cryptocurrency that fuels Ethereum transactions, which is appropriately called **Ether**.

Gas and Ether

Ethereum specializes in executing code that powers DApps. Simply put, this costs money—money to acquire the machines, to power them up, store them, and cool them if needed. That's why Ether was invented. When people talk about the price of Ethereum, they actually are referring to Ether, the currency that incentivizes people to run the Ethereum protocol on their computer. This is very similar to the way Bitcoin miners get paid for maintaining the Bitcoin blockchain.

In order to deploy a smart contract to the Ethereum platform, its author must pay to do so. That payment is made in the form of Ether. This has an interesting side effect. It motivates people to write optimized and efficient code and not waste the Ethereum network computing power on unnecessary tasks.

Ether was first distributed in Ethereum's original initial coin offering in 2014. Back then, it cost around $0.40 to buy.

Ethereum ERC-20 Token

Ethereum tokens gained popularity in 2016 and 2017 as Initial Coin Offerings or ICOs began to use them to represent ownership or value. However, the main aim of deploying Ethereum tokens was to represent in-game assets in 2017, such as in the popular game CryptoKitties. One of the most prominent aspects of tokens is their trading potentiality. For example, if you purchase a token, you could trade it for another valuable token or probably for Ether. Therefore, the standardization of tokens is crucial, which is why Ethereum Request for Comments (ERC) was created. ERC is an open and public mechanism inspired by the well-known internet Request for Comments (RFCs). RFCs allow anyone to create and comment on recommendations for defining Ethereum smart contracts and tokens.

ERC-20 and ERC-721 are two of the most widely used ERC token specifications. They are used to symbolize fungible and non-fungible assets, respectively. This section will look at the structure of **ERC-20 and ERC-721 tokens** and discuss the differentiating factors of these tokens, including how they work.

A token can be created (a.k.a. minted) by a smart contract, which is also used for managing transactions of the token and keeping track of the balance of each token holder. To purchase tokens, a buyer sends Ether to the smart contract affiliated with that token. In return, the buyer gets access to some number of tokens.

Hence minting a token means creating a smart contract that follows a set of rules specified in ERC-20 that dictates what it means to create, transfer, and keep track of account balances.

While this sounds straightforward, it can also be quite risky. As mentioned before, once a smart contract is deployed, it cannot be changed. Therefore, if you find a coding error, you cannot fix it. Imagine a bug inside your contract code that causes buyers to lose their tokens (think DAO). The one way around the stress of writing good contracts is peer reviews and testing... and lots of it.

Token Contract

Today there are a growing number of online **marketplaces** or **exchanges** in the crypto token industry that have an option to buy and sell tokens. Opensea.io is one such marketplace for the exchange of NFT tokens, which we will look at a little later.

Each token's smart contract can be completely different from another. If a developer wants their token to be available on an exchange, the exchange has to include custom code that can interoperate with contracts to allow buyers to trade. This is true for wallet providers, who today support hundreds of varieties of tokens. Building custom integration code for all permutations of tokens would be very complex and very time-consuming. This is the exact problem that the ERC-20 spec is addressing, trying to bring consistency and interoperability to the Wild West of tokens.

With specifications like ERC-20, however, exchanges and wallet providers only have to implement this code once, and that's why exchanges can add new tokens so quickly and why wallets like MyEtherWallet host hundreds of ERC-20 tokens.

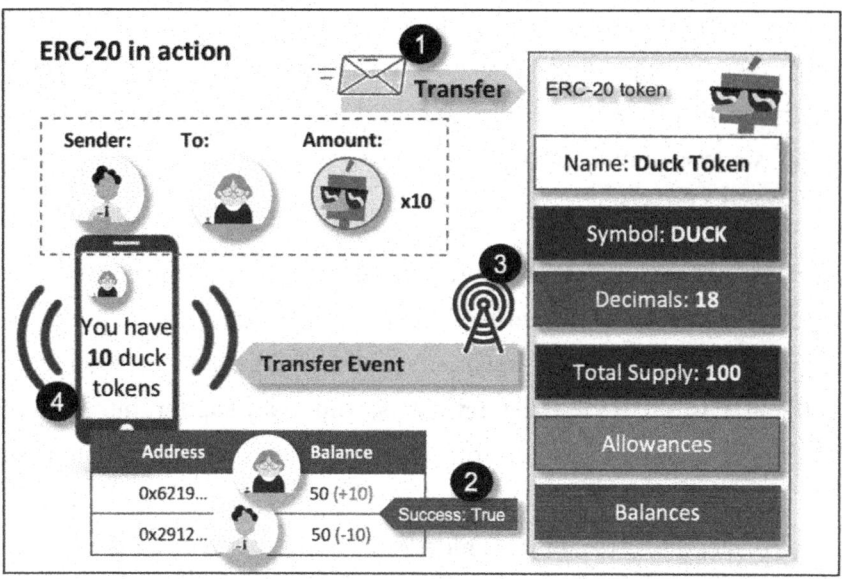

Chapter 5 – Figure 5 – Anatomy of an ERC-20 token… in action

The above figure illustrates an ERC-20 conformant exchange of a "Duck Token" between two users on the Ethereum platform. The duck token itself really doesn't do much. But we will use it here as a means to illustrate the anatomy of a token and later we will use it to code a simple smart contract version of the Duck Token in Solidity.

In the example above, one user is transferring 10 tokens to a second user. A Duck Token smart contract that conforms to ERC-20 specification would need to implement these key properties:

Property	Description	Value
Name	Defines what the token is called	**"Duck Token"**
Symbol	Think of this as the token's "ticker" symbol	**"DUCK"**
Decimals	Number of decimals determines how divisible the token is	**"18"**

Total Supply	Total number of tokens permitted to be minted	100
Allowances	Allows an address to give an allowance to another address to be able to retrieve tokens from it	
Balances	Account balances describing addresses (accounts) and the tokens assigned to their address (balance)	

These properties enable a set of standard functions that each token must support to comply, including:

Function	Description
totalSupply	Returns the current supply of tokens for the contract
BalanceOf	Returns the token balance for a specified address. In this example, balanceOf is being called with 0x6219... should return 50 tokens
transfer	Requires a recipient address as well as an amount to send. The function will only work if the amount specified is less than or equal to the sender's balance. We can see in the example above that the transfer simply decreases the sender's balance by 10 and increases the recipient's balance by 10. Once complete, the function returns a true value to show that the transfer was successful
allowance	The allowance function is used to see how much an address is allowed to spend from another address

approve	Allows a user to withdraw from an account multiple times, up to a specified amount

Lastly, ERC-20 supports a set of events that are triggered when a transaction changes a contract's state. They are important because they provide information to computing systems that are external to the Ethereum blockchain. Hence in ERC-20 token contracts, events are fired when either the Balances or Allowances are altered.

Event	Description
transfer	Triggers when tokens are transferred, including zero value transfers. In the example above, the transfer of 10 tokens between two users causes a Transfer event that notifies an external system, which ultimately has a mobile phone alert sent to the token recipient that they have 10 new Duck Tokens.
approve	Triggers when the approve function is called

NFTs with ERC-721

ERC-20 is a useful standard that has propelled the use of tokens. But it is still less than perfect. In fact, the ERC-223 standard has emerged in an attempt to make it more robust. With that, ERC-20 is designed to support fungible tokens. ERC-721, on the other hand, are Non-Fungible Tokens. Specifically, an ERC-721 token represents a class of assets, whereas an ERC-20 token represents a particular type of asset.

For example, CryptoKitties is a game where you collect and breed virtual or digital kitties. A unique ERC-721 token represents all the game's unique cats and who owns which of them.

ERC-721 simplifies ownership over ERC-20. A participant either completely owns or does not fully own an asset. For example, it is not feasible to possess "half a kitten" in CryptoKitties. Therefore, the ERC-721 token is often referred to as a standard for non-fungible assets. This is one of the crucial aspects of the ERC-721 standard to understand. However, the rest of the standards, particularly in terms of token transfers, are more or less similar to the ERC-20 standard.

CODING A SMART CONTRACT AND TOKENS

Are you ready to roll-up your sleeves to write a smart contract in the Solidity language? Well, in this section, we will write a smart contract, create an ERC-20 token, and talk about how to extend that to an NFT.

A Digital Candy Machine

I really like the vending machine metaphor for smart contracts as described by Nick Szabo. So, our example will explore a contract that represents a simple digital vending machine. As we've already discussed, the first step of creating a smart contract is the design. We need to ensure that with the right inputs, a certain output is guaranteed. Therefore, our abstract contract design dictates the following in order to get a snack from a vending machine:

> **money + snack selection = snack dispensed**

This logic now needs to be coded into our digital vending machine. Here's a simple example of how this vending machine might look like as a Solidity smart contract. The contract is only 40 lines of code, which we will look at in two parts.

```
01  pragma solidity 0.8.7;
02
03  // contract that enforces "money + snack selection = snack dispensed"
04  contract VendingMachine {
05
06      // Declare state variables of the contract
07
08      // owner of the vending machine
09      address public owner;
10
11      // account balances of candybar buyers
12      mapping (address => uint) public candybarBalances;
13
14      // When 'VendingMachine' contract is deployed:
15      // 1. set deploying address as owner of contract/vending machine
16      // 2. set deployed smart contract's candybar balance to 100
17      constructor() {
18          owner = msg.sender;
19
20          candybarBalances[address(this)] = 100;
21      }
```

You can see right away that Solidity is quite similar to JavaScript. We start the contract definition on line 4 with the keyword *contract*, specifying the **VendingMachine** object in a way similar to how we set up *class* objects in the prior chapters' examples. The VendingMachine class is simple, having two global variables, a constructor, and two methods. By the way, as we will see, this vending machine dispenses (very expensive) candy bars.

The contract defines VendingMachine **owner** and **candybarBalances** properties. As the above code snippet shows, the constructor method sets the owner to the address of the user that deployed the contract. It then goes ahead (line 20) to set the balance of candy bars (owners' inventory) in the owner's account balance to 100.

Now that we have a VendingMachine contract established, we can define the actions that are associated with the contract as seen in the following code snippet.

```
22
23      // Allow the owner to increase the smart contract's candybar balance
24      function refill(uint amount) public {
25
26          require(msg.sender == owner, "Only the owner can refill.");
27
28          // adds more candybars to owners balance
29          candybarBalances[address(this)] += amount;
30      }
31
32      // Allow anyone to purchase candybars
33      // To purchase, you must pay a least 1 ETH (expensive!)
34      // To purchase, there must be bars in stock from owner account
35      // If the above conditions are met.. we debit owner and credit buyer
36      function purchase(uint amount) public payable {
37
38          require(msg.value >= amount * 1 ether,
39                              "You must pay at least 1 ETH per candybar");
40
41          require(candybarBalances[address(this)] >= amount,
42                  "Not enough candybars in stock to complete this purchase");
43
44          // debit owners' account
45          candybarBalances[address(this)] -= amount;
46
47          // add new user to candybar balance list
48          candybarBalances[msg.sender] += amount;
59      }
50  }
```

The two methods defined above allow the VendingMachine owner to **refill** the machine, and a buyer to **purchase** a candy bar from the machine. The refill method takes an amount to refill as input and will not execute unless the refill method is being called by the owner address. The *require* keyword (on line 26) enforces this as a rule of the contract.

The purchase method also enforces rules before allowing a purchase. The first rule is to require that at least 1 Ether is paid per candy bar (as I said, these bars are expensive). The second rule (line 41) ensures the owner actually has enough bars in their inventory to support the purchase. If these conditions of the contract are met, then the accounts are debited (owner) and credited (buyer) accordingly. The buyer is now the proud owner of peanut-butter and chocolate goodness.

An ERC-20 Duck Token

I found a piece of Solidity code in Github from OpenZepplin called ERC20.sol. It implements the ERC-20 specification and makes it very easy to create new tokens. So, I took the "easy way out" to code Duck Token. However, even in this greatly simplified rendition, you can see many of the points that were illustrated above in the code that follows.

```
01  pragma solidity ^0.8.4;
02
03  // The ERC-20 spec is implemented in ERC20.sol, by importing
04  // it, we save are avoiding duplicating great deal of code here
05  import "@openzeppelin/contracts/token/ERC20/ERC20.sol";
06
07  // DuckToken is meant to be a very simple example of an ERC20 token.
08  // In the example all tokens are pre-assigned to the creator.
09  // Can later distribute these tokens using `transfer` and other ERC20 functions
10
11  contract DuckToken is ERC20 {
12
13      // "super constructor" for DuckToken, which calls on ERC20 contractor
14      // passing in token name  = "Duck Token" and symbol = "DUCK";
15
16      constructor() ERC20("Duck Token", "DUCK") {
17
18          // Mints 1,000 tokens to your wallet's address
19          _mint(msg.sender, 1000);
20      }
21  }
```

Line 5 imports the ERC20.sol code, which does much of the "heavy lifting" and is why this token only takes 21 lines to code (with comments of course). Duck Token is coded as a smart contract, hence the contract keyword is used for its definition (on line 11). The constructor method accepts a **name** and ticker **symbol**. This is a "super constructor" in that it is really delegating to the ERC20 contract constructor that is defined in ERC20.sol. This call initializes all of the ERC-20 properties that we went over in the last section. Now you can consider this part cheating, but I call it powerful. Next, we call (on line 19) the powerful **_mint** function (again, part of the ERC20 contract definition) which "under the covers" sets properties like **totalSupply** and updates a **balance** to reflect the initial balance of Duck Tokens, which is set to 1000. The _mint function also causes a **transfer** event to be emitted to notify any listeners of the newly minted Duck Tokens. That's a bit of a mouthful, but it is not really hard to create an ERC-20 token once you digest these basic concepts.

An ERC-721 Duck Token NFT

ERC-721 further extends the ERC-20 token specification by enabling the definition of unique, Non-Fungible Tokens. The primary difference is ERC-721's addition of a unique token identification number (**tokenID**) and an external (off-chain) link that references a collection of data (metadata) that represents the unique properties of this token (**tokenURI**).

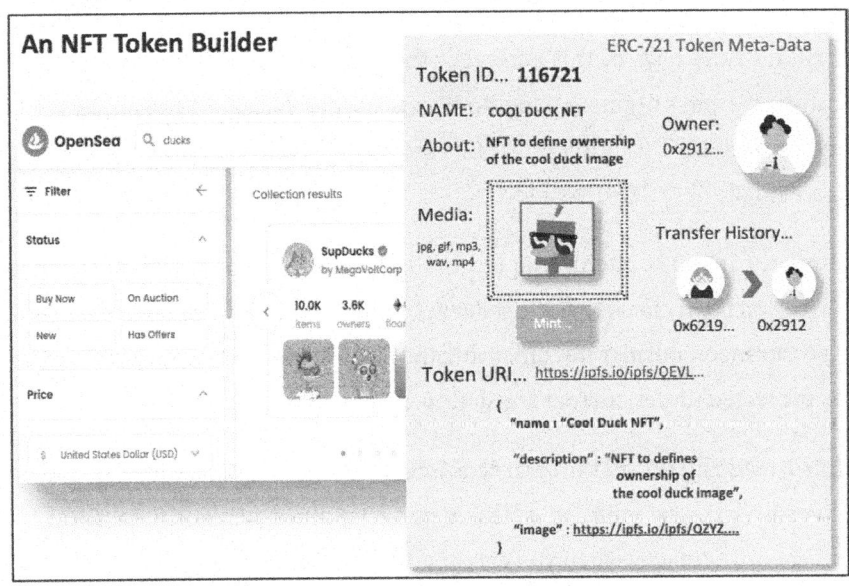

Chapter 5 – Figure 6 – Token Builders like Opensea.
io enable low-code construction of NFT tokens

There are several token builder-tools that allow for the web-based creation of ERC-20 and ERC-721 tokens without coding. Such tools as Opensea.io allow NFTs to be created, offered for sale in its marketplace and purchased. Using this tool, one can easily define the unique properties of the NFT, including references to media like images (jpgs, gifs), audio/video files (mp3/mp4), and digital documents. The NFT is given a name and description with a means to set the offering price of the NFT along with options of blockchain platforms (e.g., Ethereum) in which it will be deployed and run. If Ethereum is its destination, such tools will auto-generate the Solidity smart contract, compile it, and deploy it to the Ethereum network. All that with a simple push of a button. With that, your NFT appears as a web page and enables you to "mint" a new token or transfer ownership to another user's address.

Buyer Beware

There are many things that could go wrong when purchasing NFTs and digital tokens in general. Here are three thoughts to keep top of mind when dealing with such assets.

First, it's early days of this emergent space. Legal frameworks that protect traditional ways of purchasing don't exist yet for NFTs. For instance, if the author of an NFT fraudulently mints more copies of your "rare" digital asset, there might be little you can do.

Similarly, if you're attempting to purchase ownership or a copyright when buying an NFT, check that the seller actually owns the asset in the first place. As insurance, consider documenting the transfer of ownership in writing, so it's protected under current legislation.

Third, read and reread the smart contract code for clauses that are not widely advertised. For example, an NFT you purchase may have exorbitant royalties attached. Therefore, when you eventually sell your NFT to someone else, the original author may be owed a big portion of the sale.

The moral of this story is "let the buyer beware" and "let the seller test and retest their contract code." And for both the buyer and seller, when in doubt, read the code.

EVOLVING ETHEREUM AND COMPETITION

Ethereum has paved the way for a broad class of digital assets that go well beyond cryptocurrency. As such, Ethereum is one of the top crypto-based blockchain networks used around the world, in terms of technology, decentralization, and market capitalization. Ethereum continues to evolve and as of the writing of this chapter, Ethereum 2.0 seems imminent.

Ethereum 2.0, also known as Serenity or ETH 2.0, is an upgrade to Ethereum on several levels. Its primary objective is to increase Ethereum's capacity for transactions, reduce fees and make the network more sustainable. To accomplish this, Ethereum will change its consensus mechanism from proof of work (POW) to proof of stake (POS).[39]

Several competing technologies have also emerged during this time to provide alternatives to Ethereum with bold claims around transaction performance, energy efficiency, and "fuel" cost. These include Solana, Polkadot, Avalanche, Binance, just to name a few. Therefore, we will end this chapter with a quick look at two of these Ethereum competitors.

Solana was launched in March 2020, which mainly seeks to offer high transaction throughput. It claims to complete 65,000 transactions per second and is looking forward to constantly increasing its scalability, hoping it will be nearly doubled from its current transaction processing speed every two years.

Solana is aiming to obtain increased compatibility with the Ethereum blockchain network. For example, the introduction of "Wormhole" serves as a bridge for moving ERC-20 tokens between Ethereum and Solana efficiently.[40]

Polkadot is a blockchain network developed and launched by Parity Labs. It was released in April 2020 and has quickly gained traction. Polkadot's functionality is similar to Ethereum, it supports DApps, NFTs, and more. Polkadot seeks to create as well as develop a network of "parachains," it does not wish to restrict itself to merely a single blockchain. This might aid in providing faster and more economical transactions via the blockchain network.

Polkadot states that its popularity stemmed from its blockchain, which allows exchange of arbitrary data instead of limiting its usage to only particular data carried on traditional blockchain networks, and hence, Polkadot can be transferred across several blockchains.

- - -

What Bitcoin did for money and payments by harnessing blockchain technology, Ethereum is doing for applications of all shapes and sizes. With a built-in scripting language and distributed virtual machine, smart contracts can be built to carry out all sorts of functions without the need for a trusted third party or central authority. With this, Ethereum ushered in a new form of digital economy focused on tokens. And perhaps someday, we will be able to exchange our airline miles for hotel stays and/or cups of coffee!

QUIZ TIME

As one of the longer chapters in the book, I have tried to have these questions act as the main takeaways from the many connecting facets of tokens, Ethereum, and smart contracts.

1 – Which option best describes Nonfungible tokens?

a. Unique and non-divisible

b. Divisible and non-unique

c. Have large transaction processing fees

d. None of the above

2 – Which line best represents a smart contract?

a.   Are immutably stored on a blockchain ledger, meaning they can't be changed once deployed

b.   Not very smart, meaning they are uncompromisingly letter strict and do not factor extenuating circumstances

c.   Solidity is a smart contract language, similar to JavaScript that supports the creation of decentralized applications or DApps

d.   All of the above

3 – Which line best describes the relationship between ERC-20 and ERC-721 tokens?

a.   ERC-20 is a specification best used to define fungible tokens, where 721 is best for non-fungible

b.   Both specifications enable the minting and exchange of consistent and interoperable tokens, bringing order to the Wild West of tokens

c.   Are hosted in emerging token marketplaces and wallets like MyEtherWallet and Oceansea.io

d.   All of the above

4 – Ethereum is evolving, while other networks are emerging that aim to both leapfrog and interoperate with Ethereum

a.   True

b.   False

5 – Think Blockchain Labs– Chapter 5 Assignment

Go to the Think Blockchain Labs Github repository.

See the README.md file and search for the Chapter 5 assignment details, but the gist of the assignment is to:

a. Run the VendingMachine and Duck Token Solidity smart contracts following the instructions in the readme

b. Enhance the Duck Token from an ERC-20 token to an ERC-721 token

c. Use OpenSea.io to import your Duck Token and experiment with how one might mint a token by following detailed instructions in the readme

UP NEXT

Over the years, I've had the pleasure of meeting many of the founding fathers of Ethereum. I've attached a few photos below. The upper-left photo features Vitalik Buterin and I, who both were on a panel in 2016 at the DC Blockchain Summit. You can watch our discussion on YouTube.[41] Vitalik is the guy far left and that's me on the right side with the bow-tie, of course. The lower-right photo features Joe Lubin (Ethereum co-founder and founder of ConsenSys) and me, discussing smart contracts on a panel at the 2016 Smart Contracts Symposium. The last photo on the lower-left was taken in NYC at the DTCC Conference in 2016, featuring (right to left) Anthony Di Iorio (Ethereum co-founder), David Rutter (R3 CEO) and Blythe Masters (Digital Assets CEO and creator of the credit default swap). I am very fortunate to have spent time with these folks to appreciate their blockchain brilliance and pleasant personalities.

With Vatlik B.

Anthony Di Iorio

Joe Lubin

Chapter 5 – Figure 7 – Jerry sharing the stage with Ethereum co-founders

- - -

Moving right along. We've covered two pioneers and have one more to go. In the next chapter, we will look at a topic that is a little different. The first two pioneers, Bitcoin and Ethereum, are global networks, with thousands of nodes that are powered by blockchain technology. The third pioneer, Hyperledger Fabric, is a plug-and-play framework by which a wide variety of blockchain networks can be written. Hence, you might say that while Bitcoin and Ethereum possess kinetic energy, Hyperledger Fabric possesses potential energy that can be brought to bear to solve an even wider variety of distributed solutions across multiple industries.

Perhaps a bit of a warning is warranted before we start this next chapter. As a framework for building arbitrary styles of blockchain, Fabric's appeal is to a network architect versus an end-user like the other chapters before this. Hence, this chapter may feel a bit more technical and difficult to read than the others in the book. And if it does, feel free to just read the history section and perhaps you can skip right down to the use cases that conclude the chapter.

Chapter 6 –
ENTERPRISE, HYPERLEDGER FABRIC, MODULARITY

Fabric is like a Lego kit that allows you to assemble the pieces to produce the Millennium Falcon from Star Wars, or a model of the Eiffel Tower. Fabric has the potential energy to become just about any type of blockchain network, with "some assembly required."

COVERED IN THIS CHAPTER

- How Hyperledger happened

- How Fabric addresses the needs of the enterprise

- Modularity enables fit-for-purpose blockchain networks

- Breakthrough in Identity, Privacy, and Consensus

- Examples of enterprise blockchains beyond cryptocurrency

HISTORY OF HYPERLEDGER (ACCORDING TO JERRY)

Hyperledger Fabric, or what many people simply refer to as Fabric, is a technology near and dear to me. So, let the reader beware that I'm a bit biased on this particular topic. I rank Fabric up there with the who's who of blockchain pioneers because of its versatility in supporting the "other 999 applications" that would benefit from blockchain's properties. I can't resist starting this chapter off with a semi-brief narration of how Hyperledger came to be. Here goes.

I was there in February 2016 when approximately 30 collaborating companies launched the Hyperledger Project under the Linux Foundation.[42] This means there are possibly 29 other versions of the history of Hyperledger. What follows here is *my* rendition of what led to the Hyperledger project along with some commentary on why I feel it stands alongside the other pioneers of blockchain.

It all starts off with an interest that IBM and I have on the subject of transaction processing. As you read this, billions of transactions flow daily through IBM hardware and software systems. In fact, IBM has brought to market some of the pioneering technologies in transaction processing. This includes software like CICS that's powering the mainframe era of transactions using a centralized, vertical scale architecture. From this architecture, we saw the birth of the world's most reliable and highest performing applications including bank payments and airline reservation systems.

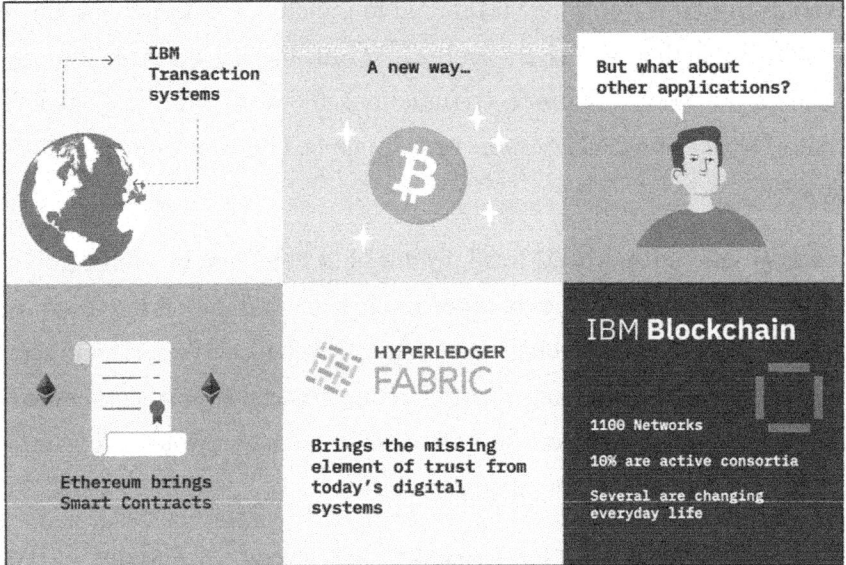

Chapter 6 – Figure 1 – Brief history of my history with Hyperledger and Fabric

Later, I was fortunate to be one of the early contributors to IBM WebSphere, which ushered in the era of web-based transaction processing that scaled with a centralized, horizontal architecture. From this we saw the rise of auction systems from eBay, online banking, and retail shopping applications that were the poster children for modern, interactive Web 2.0 apps. And while web-based processing led to cloud and then mobile, the centralized scale-out architecture prevailed. I started to wonder if there would ever be a next chapter or was Web2 as good as it was going to get for transaction processing.

In 2013, our work on WebSphere expanded out beyond web to mobile applications. My team was tasked with looking for opportunities beyond standard mobile development and we quickly started exploring mobile payments. Since IBM is not a bank, we looked for adjacencies to payments and landed on the interesting topic of gamification. There are many examples of gamification, the most well-known perhaps being frequent flyer rewards programs offered by airlines. As we discussed in the previous chapter, frequent flyer rewards are a type of reward token. Studying the token economy and modern technology that would drive it, we discovered Bitcoin, and then Ethereum.

Friends from the WebSphere team used to mine Bitcoin during our lunch breaks. While this may have been the first time blockchain made it to my radar as "a new way" to process transactions, it's wasn't clear how Bitcoin's style of blockchain could be more broadly applied to "all the other applications".

It was around 2014, when I was introduced to Ethereum by Binh Nguyen and Frank Lu, two members of my technology team. That was it! I instantly fell in love with the technology, especially blockchain and smart contracts. "This is the next game-changer in transaction processing" is what I shouted. Simply put, smart contracts were the missing piece that would allow us to apply blockchain to just about every industry.

Now, IBM is a big place, it just so happens that several other groups in IBM were coming to this realization from different angles. These teams joined forces, and this is where my blockchain journey started. Throughout 2014, we studied Ethereum, and even talked to and met some of the key contributors within the early Ethereum ecosystem. As we gained a broader understanding of what made Ethereum tick, we started to realize that a pseudonymous, completely decentralized, crypto-incentivized architecture would not be easy to consume by many of our enterprise customers. Specifically, these customers participated in regulated industries, and depended on governments to keep commerce fair, hence anonymous and completely decentralized architectures was not a good place to start a conversation.

But Ethereum was a good place to be in 2014, and we went on to have talks with the Ethereum community to extend it with a permissions framework. Our proposal for a "Permissioned Ethereum" received a lot of interest, but our timing was off because the Ethereum world was about to launch their main net. The poor timing was exacerbated by the fact that the various Ethereum codebases each had their own licensing agreements; none were particularly viewed as inviting by our IBM lawyers (e.g., GPL, LGPL).

It was also true that our participation wasn't welcomed by some (a small minority actually, but it was distracting no less). Our views were minimized by some saying, "You really don't know what you're talking about because you've never built a blockchain."

So, we decided to build one, if not for any other reason than to learn with our hands. We also realized that we should not do this in isolation and that open source was the only way to go. The Linux foundation has always been the gold standard for open-source projects and governance, so of course we went with them. John Wolpert, Chris Ferris, and I met with Jim Zemlin from the Linux Foundation to pitch our project that we were calling Open Block Chain (OBC). Jim liked the concept and informed us that companies had approached him with similar interests. With Jim's team's help, in December of 2015 (announced in early 2016), we were among the 30 companies that launched our open blockchain initiative. One of the companies, Digital Assets, had acquired a small company called Hyperledger, and they generously donated the (domain) name. IBM also donated about 50,000 lines of some of our early (OBC) code, which ultimately became the Hyperledger Fabric sub-project.

Today, Hyperledger Foundation hosts several enterprise-grade blockchain software projects. The projects are conceived and built by the developer community for vendors, end-user organizations, service providers, start-ups, academics, and others to use to build and deploy blockchain networks or commercial solutions.

If contribution and adoption are the key measurements of success, then the Hyperledger Fabric project quickly stood out as an early leader. To this day, IBM and a worldwide list of contributors continue to grow and expand the Fabric codebase. Several companies, including IBM, Amazon, and Kaleido, offer commercial support for the technology. This culminates in a compelling list of network creators adopting Hyperledger Fabric for their projects. We have tracked over eleven hundred blockchain networks built atop Fabric to date, which span across all industry use cases, fulfilling the order of a blockchain for enterprise.

Now that I've shared the history and motivation behind Hyperledger, let's proceed using the approach used in the previous chapters. Specifically, looking at this topic from three angles; target application set (i.e., Enterprise), the blockchain technology that pioneered this style of enterprise use cases (i.e., Hyperledger Fabric), and how it accomplished it (i.e., Modularity).

A BLOCKCHAIN FOR ENTERPRISE

The broader application of blockchain, beyond cryptocurrency and tokens is what often gets referred to as blockchain for business or simply, enterprise blockchain. The three stories covered in Chapter 2 are my favorite examples of not only applying blockchain broadly, but they are also great examples of how blockchain can be used to stamp out some real impediments that are faced by digital businesses, including ownership disputes, identity theft, and counterfeiting. In this chapter, we will gain a better understanding of why these examples use Hyperledger Fabric.

Fabric is a modular, general-purpose framework, whose unique features make it suitable for a variety of industry applications such as track-and-trace of supply chains, trade finance, loyalty and rewards, as well as clearing and settlement of financial assets.

For enterprise use, we need to consider the following requirements:

- Participants must be identified/identifiable

- Networks need to be *permissioned* such that members are known

- High transaction throughput performance

- Low latency of transaction confirmation

- Privacy and confidentiality of transactions and data pertaining to business transactions

While many early blockchain platforms are currently being adapted for enterprise use, Fabric has been *designed* for enterprise use from the outset. What follows is a description of how Fabric differentiates itself from other blockchain platforms and describes some of the motivation for its architectural decisions.

Open

As described in the history section earlier, Hyperledger Fabric is an **open source** blockchain established under the Linux Foundation, which itself has a long and very successful history of nurturing open source projects under **open governance** that grow strong, sustaining communities and thriving ecosystems. Hyperledger is governed by a diverse technical steering committee, and the Hyperledger Fabric project by a diverse set of maintainers from multiple organizations. It has a development community that has grown to over 35 organizations and nearly 500+ developers since its earliest commits.

Modular

Fabric has a highly modular and configurable architecture, enabling innovation through experimentation, versatility, and optimization for a broad range of industry use cases including banking, finance, insurance, healthcare, human resources, supply chain and even digital music delivery. Many of Fabric's capabilities described throughout this chapter are "modules" that contribute to the plug and play nature of Fabric.

As a modular blockchain development toolkit, the prototypical users of Fabric are **network creators**. This is different from Bitcoin and Ethereum, whose user persona is more of a **network end-user**. To further dramatize this point, if a toolkit like Fabric existed earlier, would network creators like Satoshi or Vitalik have used it to create their blockchain masterpieces? While fun to ponder, the point here is more to stress that Fabric modularity really sets the sky as the limit when it comes to the types of blockchain networks that can be created.

General Purpose & Programmable

Fabric is the first distributed ledger platform to support smart contracts authored in general-purpose programming languages such as Java, Go and Node.js, rather than constrained Domain-Specific Languages (DSL). This means that most enterprises already have the skill set needed to develop smart contracts, and no additional training to learn a new language or DSL is needed.

Permissioned

The Fabric platform is often referred to as 'permissioned', meaning the participants are known to each other, rather than anonymous. This means that while the participants may not *fully* trust one another (they may, for example, be competitors in the same industry), a network can be operated under a governance model that is built off whatever trust *does* exist between participants, such as a legal agreement or framework for handling disputes.

Given Fabric's modular design, different styles of network membership and access control can be developed, which can alter the behavior of how users are identified by the network. It is even conceivable that a permissionless architecture can be developed within Fabric's framework. With that and given there are many permissionless frameworks available on the market, Fabric is almost exclusively used for permissioned use cases.

Pluggable Consensus

One of the standout features of the platform is its support for pluggable consensus protocols that enable it to be more effectively customized to fit particular use cases and trust models. For instance, when deployed within a single enterprise, or operated by a trusted authority, fully Byzantine Fault Tolerant (BFT) consensus might be considered unnecessary and an excessive drag on performance and throughput. In situations such as that, a Crash Fault Tolerant (CFT) consensus protocol might be more than adequate whereas, in a multi-party, decentralized use case, a more traditional BFT consensus protocol might be required. BFT assumes network participant to be untrustworthy, which might be a reasonable assumption if competitors are on the same network. CFT is more trusting, however, provisions need to be taken if nodes unintentionally fail (i.e., crash). There is a section to follow that explains BFT in more detail. So, if this is a new concept to you, please just hang in there and it will be explained in a minute.

Fabric can leverage consensus protocols that **do not require a native cryptocurrency** to incent costly mining or to fuel smart contract execution. Avoidance of a cryptocurrency reduces some significant risk/attack vectors, and absence of cryptographic mining operations means that the platform can be deployed with roughly the same operational cost as any other distributed system.

The combination of these differentiating design features makes Fabric one of the better performing platforms available today both in terms of transaction processing and transaction confirmation latency, and it enables privacy and confidentiality of transactions and the smart contracts that implement them.

Let's explore these features in more detail.

THE LEGO OF BLOCKCHAIN

Bitcoin and Ethereum are purposed cryptocurrency networks that are pre-assembled, always running, and operate 24 x 7. Fabric is quite different in that it is first and foremost a modular technology framework for building a wide variety of blockchain networks. We like to use the analogy that Fabric is like a Lego kit that allows you to assemble the pieces to produce the Millennium

Falcon from *Star Wars,* or a model of the Eiffel Tower. Fabric has the potential energy to become just about any type of blockchain network, with "some assembly required."

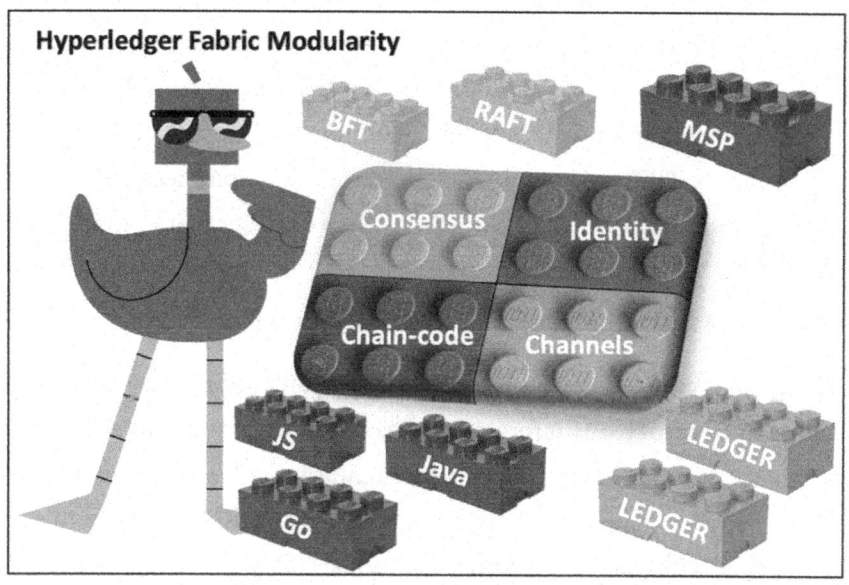

*Chapter 6 – Figure 2 – Fabric's "Plug and Play" architecture enables
a wide variety of blockchain use case exploration.*

As the above figure highlights and as the following sections will detail, Fabric's architecture enables four areas of extensibility. While Fabric enables new plug-ins to be built and experimented with in these areas, it also provides a growing list of already-built options that specifically address enterprise concerns about accountability, privacy, performance, and security.

IDENTITY WITH CONFIDENTIALITY

Identity – Fabric's membership service provider (MSP) abstracts away all cryptographic mechanisms and protocols behind issuing certificates, validating certificates, and user authentication. This is the key to supporting enterprise requirements for accountability and privacy. Known participation need not mean your dirty laundry can be seen by all. This is where known identity meets confidentiality.

Administrators, end-users and blockchain network components (e.g., nodes) must first all register and enroll to receive a "membership card" (i.e., certificates) to perform their role in a Fabric blockchain network. This can be viewed similar to receiving your "employee badge" that allows access to the various buildings of your company, while restricting access to only the parts of the building that you have permission to use. The enrollment process is the foundation enabling participants to be known to the network. However, not everyone has a need to know. This is where Fabric's confidentiality features kick in and complement identity.

As we have discussed, in a public, permissionless blockchain network that leverages POW for its consensus model, transactions are executed on every node. This means that neither can there be confidentiality of the contracts themselves, nor of the transaction data that they process. Every transaction, and the code that implements it, is visible to every node in the network. In this case, we have traded confidentiality of contract and data for a public consensus model, operated by pseudonymous miners.

This lack of confidentiality can be problematic for many business/enterprise use cases. For example, in a network of supply-chain partners, some consumers might be given preferred rates as a means of either solidifying a relationship or promoting additional sales. If every participant can see every contract and transaction, it becomes impossible to maintain such business relationships in a completely transparent network—everyone will want the preferred rates!

As a second example, consider the securities industry, where a trader building a position (or disposing of one) would not want her competitors to know of this, or else they will seek to get in on the game, weakening the trader's gambit.

In order to address the lack of privacy and confidentiality for purposes of delivering on enterprise use case requirements, blockchain platforms have adopted a variety of approaches. All have their trade-offs. Encrypting data is one approach to providing confidentiality; however, in a permissionless network leveraging POW for its consensus, the encrypted data is sitting on every node. Given enough time and computational resource, the encryption could be broken. Considering the fact that some financial instruments can take ten years or more to come to value, the risk of cryptography breaks over time (i.e., think quantum computers) could allow private information to become public. Therefore, for many enterprise use cases, the risk that their information could become compromised is unacceptable.

Zero-Knowledge Proofs (ZKP) are another area of research being explored to address this problem, the trade-off here being that, presently, computing a ZKP requires considerable time and computational resources. Hence, the trade-off in this case is performance for confidentiality.

In a permissioned context that can leverage alternate forms of consensus, one might explore approaches that restrict the distribution of confidential information exclusively to authorized nodes. Hyperledger Fabric uniquely enables confidentiality through its channel architecture and private data features, which are covered in the following section.

CHANNELS FOR PRIVACY

Channels – Participants on a Fabric network establish a sub-network, where every member has visibility to a particular set of transactions. Thus, only those nodes that participate in a channel have access to the smart contract and data transacted, preserving the privacy and confidentiality of both. Private data allows collections between members on a channel, allowing much of the same protection as channels without the maintenance overhead of creating and maintaining a separate channel.

Perhaps the easiest way to understand Fabric's channel design is to liken it to channels in Slack. Slack organizes conversations into dedicated spaces (also) called channels. Slack Channels bring the right people and information together for a common purpose—like a specific project, topic, or team. Content shared in a channel is private to the members that have been invited to participate. The data is inaccessible to others in the network.

A Fabric channel comes with its own individualized ledger, which is the blockchain transaction log for all transactions that occur within the context of that channel. So, in essence, a Fabric network comprises a collection of blockchain ledgers, versus a single monolithic ledger as seen in many other blockchain architectures. And for those who study database technology, a channel can also be viewed as a shard, or data partition. Fabric supports a modular ledger design that allows for different data storage technology to be plugged-in as a ledger store. For example, Fabric currently supports plug-ins for Level-DB and CouchDB.

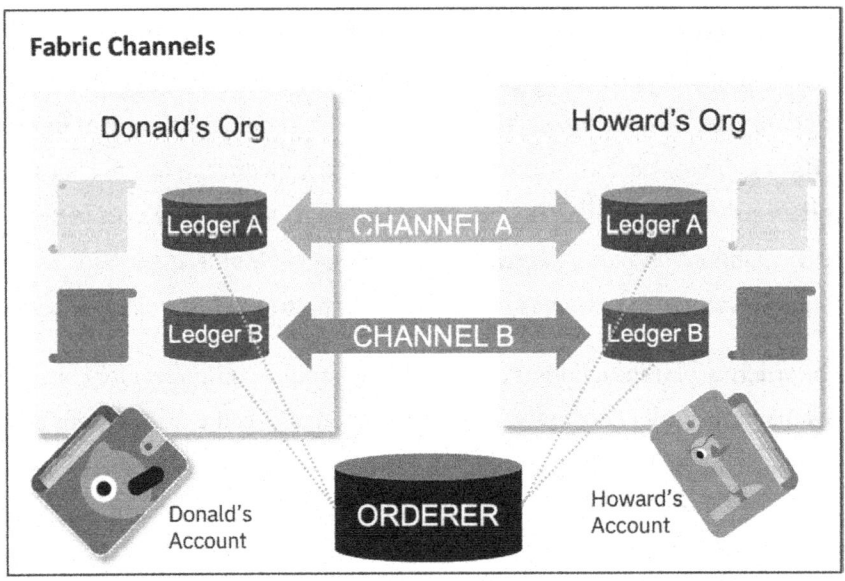

Chapter 6 – Figure 3 – Relationship between Fabric orgs, ledgers, orderers and channels

As the figure above shows more specifically, a channel is defined by members (organizations), anchor peers per member, the shared ledger, smart contracts, and the ordering service node(s). Each transaction on the network is executed on a channel, where each party must be authenticated and authorized to transact on that channel. Each peer that joins a channel has its own identity given by a Membership Services Provider (MSP), which authenticates each peer to its channel peers and services.

Simply put, the combination of identity (gained through the MSP and enrollment process) and confidentiality and privacy (gained through channels) is the equivalent of "having your cake and eating it too." In other words, this combination takes a giant step towards making blockchain usable in many enterprise contexts. However, two more breakthroughs were needed as you will see explained in the next two sections on consensus and smart contracts.

CONSENSUS - A UNIQUE APPROACH

Consensus – Plug-able consensus means fit-for-purpose algorithms can be employed to accommodate higher performance and availability models while validating blockchain transactions. It features a flexible endorsement model for achieving consensus. It is designed to allow network starters to choose a consensus mechanism that best represents the relationships that exist between participants across required organizations. A pluggable ordering service establishes consensus on the order of transactions and then broadcasts blocks to peers.

The ordering of transactions is delegated to a modular component for consensus that is logically decoupled from the peers that execute transactions and maintain the ledger. Specifically, the **ordering service**. Since consensus is modular, its implementation can be tailored to the trust assumption of a particular deployment or solution. This modular architecture allows the platform to rely on well-established toolkits for CFT (Crash Fault-Tolerant) or BFT (Byzantine Fault-Tolerant) ordering.

Fabric introduces a new architecture for transactions that is call **execute-or-der-validate**. It addresses the resiliency, flexibility, scalability, performance and confidentiality challenges faced by the order-execute model by separating the transaction flow into three steps:

- *execute* a transaction and check its correctness, thereby endorsing it,

- *order* transactions via a (pluggable) consensus protocol, and

- *validate* transactions against an application-specific endorsement policy before committing them to the ledger

This design departs radically from the order-execute paradigm in that Fabric executes transactions before reaching final agreement on their order.

In Fabric, an application-specific endorsement policy specifies which peer nodes, or how many of them, need to vouch for the correct execution of a given smart contract. Thus, each transaction need only be executed (endorsed) by the subset of the peer nodes necessary to satisfy the transaction's endorsement policy. This allows for parallel execution increasing overall performance and scale of the system. This first phase also **eliminates any non-determinism**, as inconsistent results can be filtered out before ordering.

Because Fabric has eliminated non-determinism, it is the first blockchain technology that **enables use of standard programming languages**. This is the topic of the next section on chain-code. However, I have promised a short explanation on what BFT consensus is all about. So, before we finish consensus, here is a quick summary.

Byzantine Fault Tolerance

The Byzantine Generals Problem was first introduced in a computer science paper published in 1982. Here is a short excerpt from the paper.

We imagine that several divisions of the Byzantine army are camped outside an enemy city, each division commanded by its own general. The generals can communicate with one another only by messenger. After observing the enemy, they must decide upon a common plan of action.

A distributed computer system, like blockchain, must be able to function effectively in the presence of faulty components that may send conflicting information to different parts of the system. This analogy fits distributed computation because individual machines (generals) are connected only via the network (messengers). Each is only aware of what it observes and the messages it receives. Likewise, every component is a possible source of failure or malicious activity.

Byzantine General	Blockchain
Messengers captured or killed	Networks can fail
Messengers compromised	Man-in-the-middle attacks can send forged messages
Generals killed	Hardware or software components can break or crash
Generals become traders	Malicious components can send malicious messages

The analogy also fits the topic of consensus because the generals must act together. The army only succeeds if its divisions act together. Either all divisions attack or all divisions retreat.

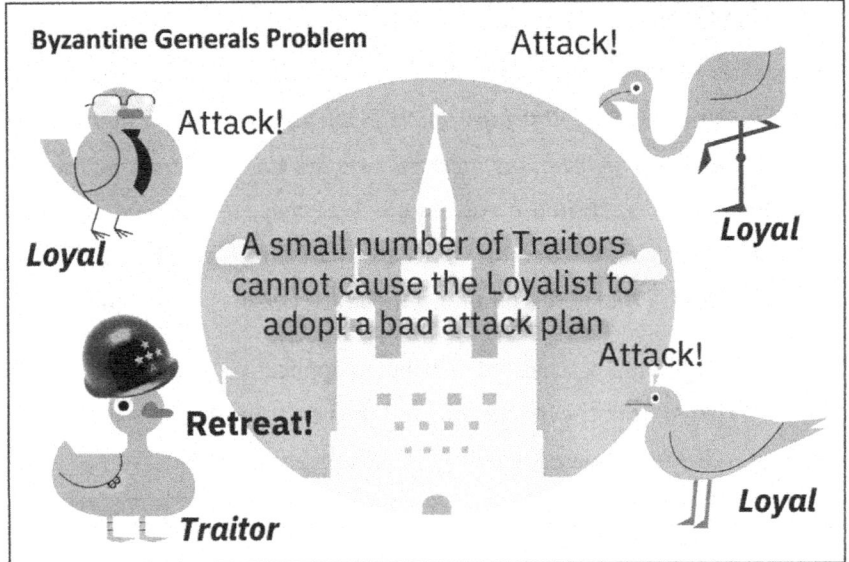

Chapter 6 – Figure 4 – Byzantine Generals Problem motivates
consensus in the presence of bad or faulty actors

If only some divisions attack, the city can defeat them, and the retreating divisions will be too weak to resist being chased down. In a blockchain, consensus is also paramount. The power of the blockchain is that it is a shared source of truth. If different parts of the network have different blockchain states, they can no longer work together—they're operating in different worlds.

With Bitcoin, Satoshi Nakomoto provided a novel solution to the Byzantine Generals Problem. Satoshi also unknowingly opened the door to new ground in consensus protocol research. While POW, POS, and BFT are definitely the most popular consensus mechanisms, many other consensus mechanisms like Delegated Proof of Stake (DPOS), Proof of Elapsed Time (PoET), Proof of History, and Directed Acyclic Graphs (DAG), among others, are emerging, which cater to the issue of BFT albeit with varied levels of success. We are yet to have a perfect consensus mechanism; however, the modular nature of Fabric invites the exploration of newer consensus algorithms in the future.

CHAIN-CODE AS SMART CONTRACTS

Chain-code – *Fabric's chain-code abstraction enables smart contracts to be defined in any number of (Touring Complete) languages. Hence, it supports the creation of business logic in languages that are known to enterprises, thereby allowing them to use in-house skills to quickly engage blockchain use cases. Specifically, Fabric supports smart contracts written in chain-code, which can be implemented in these programming languages; Go, Java, and Node.*

Chain-code functions as a trusted distributed application (DApps) that gains its security/trust from the blockchain and the underlying consensus among the peers. It is the business logic of a blockchain application.

There are three key points that apply to smart contracts:

- many smart contracts run concurrently in the network

- they may be deployed dynamically (in many cases, by anyone), and

- application code should be treated as untrusted, potentially even malicious.

Most existing smart-contract capable blockchain platforms follow an **order-execute** architecture in which the consensus protocol:

- validates and orders transactions, then propagates them (in blocks) to all peer nodes,

- each peer then executes the transactions sequentially.

The order-execute architecture can be found in virtually all existing block-chain systems, ranging from public/permissionless platforms such as Ethereum (with POW-based consensus) to permissioned platforms such as Tendermint, Chain, and Quorum.

Smart contracts executing in a blockchain that operates with the order-execute architecture must be deterministic; otherwise, consensus might never be reached. To address the non-determinism issue, many platforms require that the smart contracts be written in a non-standard, or domain-specific language (such as Solidity) so that non-deterministic operations can be eliminated. This hinders widespread adoption because it requires developers writing smart contracts to learn a new language and may lead to programming errors.

Further, since all transactions are executed sequentially by all nodes, performance and scale are limited. The fact that the smart contract code executes on every node in the system demands that complex measures be taken to protect the overall system from potentially malicious contracts in order to ensure resiliency of the overall system.

The next section looks at how these four modules of Fabric (identity, channels, consensus, and chain-code) come together to process blockchain transactions while maintaining enterprise qualities.

FABRIC IN ACTION

Fabric's modular architecture separates transaction processing into a workflow with three specific stages:

Stage 1	Chain-code that comprise the distributed logic processing
Stage 2	Transaction ordering
Stage 3	Transaction validation and commitment.

This segregation offers multiple benefits:

- Reduces verification steps

- Improved network scalability

- Better overall performance

Additionally, Fabric's support for plug-and-play of various components allows for easy reuse of existing features and readymade integration of various modules. For instance, if a function already exists that verifies the participant's identity, an enterprise-level network simply needs to plug-in and reuse this existing module instead of building the same function from scratch.

The participants on the network have two primary roles:

Endorser - The transaction proposal is submitted to the endorser peer according to the predefined endorsement policy that establishes the number of endorsers required. After sufficient endorsements by the endorser(s), a batch or block of transactions is delivered to the committer(s).

Committer - Committers validate that the endorsement policy was followed and that there are no conflicting transactions. Once both the checks are made, the transactions are committed to the ledger.

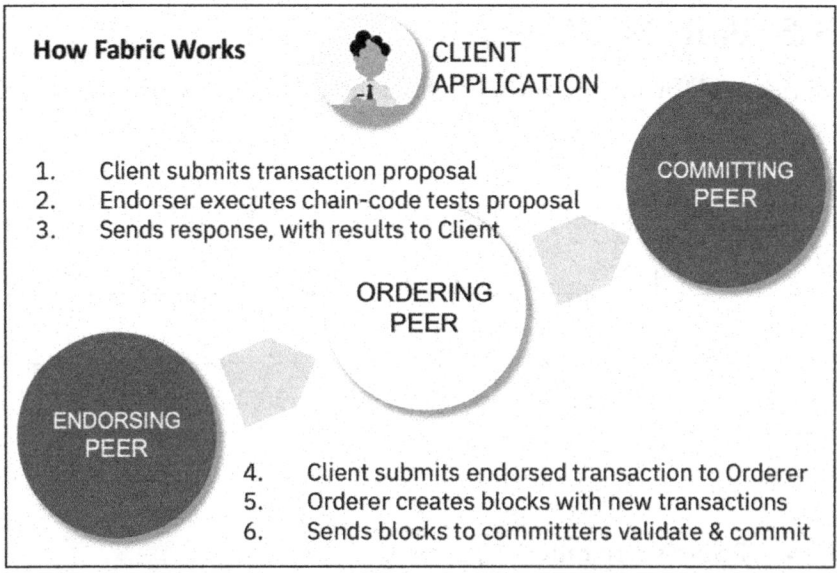

Chapter 6 – Figure 5 – How Fabric Works

Since only confirming instructions, such as endorsement signatures, are sent across the network, the scalability and performance of the network is enhanced. Only endorsers and committers have access to the transaction, and security is improved with a fewer number of participants having access to key data points.

An Example

Suppose there's a manufacturer that wants to ship ducks to a specific retailer or market of retailers (i.e., all US retailers) at a specific price but does not want to reveal that price in other markets (i.e., Canadian retailers).

Since the movement of the product may involve other parties, like customs, a shipping company, and a financing bank, the private price may be revealed to all involved parties if a basic version of blockchain technology is used to support this transaction.

Fabric addresses this issue by keeping private transactions private on the network; only participants who need to know are aware of the necessary details. Data partitioning on the blockchain allows specific data points to be accessible only to the parties who need to know.[43]

With a unique approach to smart contracts, consensus, channels for privacy and identity with confidentiality, Fabric networks can get right down to business and perform. The next section briefly comments on a critical enterprise feature that falls out from Fabric's design, which is performance and scale.

PERFORMANCE AND SCALABILITY

Performance of a blockchain platform can be affected by many variables such as transaction size, block size, network size, as well as limits of the hardware, etc. The Hyperledger Fabric Performance and Scale working group currently works on a benchmarking framework called Hyperledger Caliper.

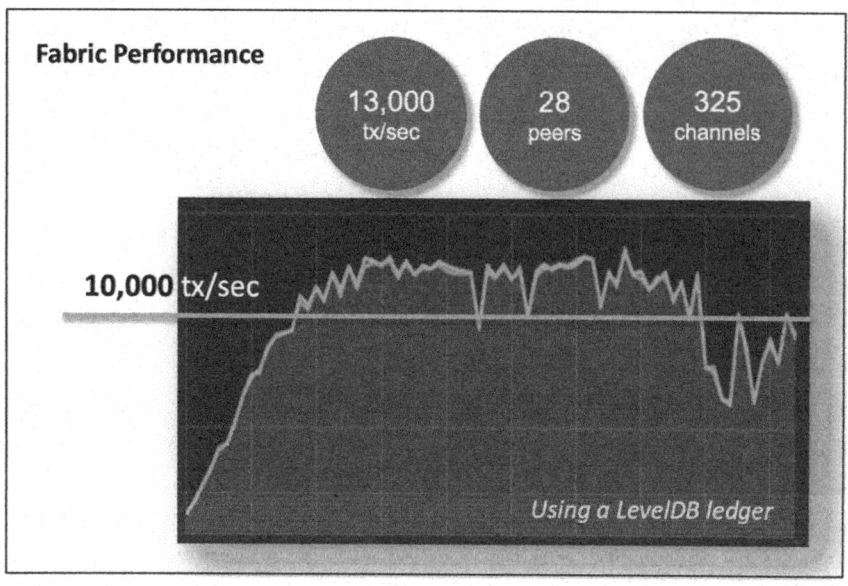

Chapter 6 – Figure 6 – An excerpt from a performance report by Chris
Ferris showing Fabric achieving over 10,000 transactions/sec

Several research papers have been published studying and testing the performance capabilities of Fabric. The figure above is an excerpt from a performance report by Chris Ferris titled, "Does Hyperledger Fabric perform at scale?" In the paper, Chris observes that:

With a system under test, including 128 peers, 325 channels using LevelDB ledger, approximately 13,000 transactions per second were achieved. Considering that each peer is processing 25 channels at a transaction rate approaching the rate of a peer running a single channel (13,000/325 * 25 = 1,000), we conclude that there is no meaningful performance degradation of the peer itself based on having 25 channels versus one.[44]

Other papers published confirmed Chris' findings with results showing Fabric scaling to 20,000 transactions per second.[45]

INDUSTRY USE CASES FOR FABRIC

Supply Chain

Supply chains are global, distributed webs of suppliers, manufacturers, and retailers. Fabric networks can improve supply chain processes by increasing transparency and traceability of transactions within the network. On a Fabric network, companies with access to the ledger can view the same immutable data, which enforces accountability and reduces the risk for counterfeiting. In addition, production updates are added to the ledger in real time, which makes tracking provenance faster and simpler during events like product recalls or food contamination outbreaks.

Fabric provides a standard protocol to allow every participant on a supply chain network to input and track the number of goods and supplies that are produced and used on a specific medium. The technology enables buyers to access a complete record of information and trust that the information is accurate and complete. This blockchain project also follows a modern approach to seafood traceability.

Trading and Asset Transfer

Trading requires many organizations such as importers, exporters, banks, shipping companies, and customs departments, to work with one another. Using Fabric, financial and trading consortiums can easily create a blockchain network, where all parties can transact and process trade-related paperwork electronically, without the need for a central trusted authority. Unlike other processes that require trade-related paperwork to go back and forth between the stakeholders, taking 5-10 days to complete, transactions in a Fabric network can process instantly.

Insurance

Insurance fraud costs the insurance industry billions of dollars a year, but with Fabric, insurance companies can reference transaction data stored on the ledger to identify duplicate or falsified claims. Blockchain can also make multi-party subrogation claims processing faster by using smart contracts to automate repayment from the at-fault party back to the insurance company. In addition, insurers can use Fabric to streamline Know Your Customer (KYC) processes by storing customer data on a distributed ledger and automating the verification of their identity documents with smart contracts.

Digital Identity

Despite security governance, security is still hampered by numerous mechanisms such as data breaching and cybercrimes that act as milestones. Fabric can help in providing robust security and reliability, thus safeguarding digital identity. It allows organizations and individuals to have complete control over their digital credentials as one can govern when, where, and with whom to share their crucial information. This topic will be covered more in the cyber-security chapter that follows.

Real estate Transactions

Fabric can help in tackling real estate transactions as well. With decentralizing databases, it can keep track of land titles, and every transaction involved would be recorded on the public ledger that will help financial firms in making decisions like, for example, whether to extend the loan to someone looking to buy a property. Further, this permissioned-based, shared system of record will increase overall trust and protect homeowners from making wrong decisions.

Music and Media Rights

Fabric creates a fair and transparent mechanism in the music and media industry. It offers a transparent method for music composers, artists, rights holders, etc., to express their rights for commercializing their art, where data is maintained across a distributed network. The technology enables seamless data exchanges between artists and users, without the involvement of any third party.

TRY FABRIC – YOU'LL LIKE IT!

Any serious evaluation of blockchain platforms should include Hyperledger Fabric in its shortlist. Combined, the differentiating capabilities of Fabric make it a highly scalable system for permissioned blockchains supporting flexible trust assumptions that enable the platform to support a wide range of industry use cases ranging from government to finance, to supply-chain logistics, to healthcare and so much more.

Blockchain-as-a-service (BaaS) providers have emerged making evaluating and running blockchain frameworks, like Fabric, a walk-in-the-park. Of these, the BaaS from Kaleido has quickly emerged as a leader in blockchain-based cloud services. (And I'm not just saying that because the founders are my friends). The Kaleido platform offers a simple experience to build and manage blockchains running across Microsoft Azure and Amazon Web Services. In fact, as you will see, this chapter's Lab Assignment features Kaleido.

QUIZ TIME

As mentioned in the last Up Next section, this chapter is a bit more technical than some of the others in the book. I hope that even if you found it pushing the limits, you still have gained an appreciation of why I consider Fabric as a blockchain pioneer with unique appeal to enterprises exploring a broader application of blockchain. And if I pushed it too far, all I can say is "b.d.d.a."

1 – Which best describes the enterprises requirements that influenced fabric's design?

    a.   Participants/Private, Networks/Open, Transaction Performance/ High

    b.   Participants/Known, Networks/Permissioned, Transaction Performance/High

    c.   Participants/Bots, Networks/Interoperable, Transaction Performance/Based on Miners

    d.   None of the above

2 – The MSP module allows registration and network enrollment for:

    a.   Client Applications

    b.   Network componentry (Peers and Ordering Nodes)

    c.   Network Administrators

    d.   All of the above

3 – Fabric networks don't require mining or cryptocurrency for fueling consensus and instead…

    a.   Enable well-established toolkits to be plugged in for CFT (Crash Fault-Tolerant) or BFT (Byzantine Fault-Tolerant) ordering

    b.   Employs a unique execute-order-validate architecture that helps address the resiliency, flexibility, scalability, performance, and confidentiality requirements

c. Because Fabric eliminates non-determinism, it's the first block-chain that enables use of standard programming languages for smart contracts.

d. All of the above

4 – Fabric's modular architecture separates transaction processing into a workflow with three specific stages; process chain-code, order transactions, validate/commit transaction

a. True

b. False

5 – Think Blockchain Labs– Chapter 6 Assignment

Go to the **Think Blockchain Labs** Github repository. See the README.md file and search for the Chapter 6 assignment details, but the gist of the assignment is to:

a. Stand-up a Hyperledger Fabric blockchain network using the **Kaleido.io** blockchain cloud service by following instructions specified in the readme.

b. Run any one of the Hyperledger Fabric chain-code samples.

UP NEXT

As I shared in the beginning of this chapter, the Hyperledger topic is near and dear to me because I've had the pleasure of leading the IBM Blockchain team for over 5 years even as much of the technology discussed here was created and brought to market. And to celebrate that, I cannot resist including a photo that the team took in 2019 "jumping for joy" at our Think Conference in Las Vegas. Also pictured (lower-right) is my friend and colleague Marie Wieck. Marie and I have been on many technology adventures over the years. When this photo was taken, Marie was leading the overall blockchain mission at IBM as our general manager.

Chapter 6 – Figure 7 –Collage of IBM Blockchain team circa 2019

One of the more memorable and rewarding parts about working on blockchain was traveling the world with colleagues collaborating with enterprises of all sizes and unlocking the potential of enterprise blockchain. In 2019, I had the pleasure of traveling to Thailand, not once, but three times that year. The above-lower-left photo is of my buddies, including Heran Shah, in Thailand, visiting with a bank consortium via crowded subway… the good old days. As seen in the following photo, I also spoke at our customer conference that day.

Chapter 6 – Figure 8 – IBM Blockchain Journey presented at Think Thailand 2019

– – –

Let's take a quick inventory of the ground we've covered thus far. We've defined blockchain in the broadest sense and then teased out the characteristics that make a permissioned blockchain different from a permissionless one. To study this further, we picked three pioneers of blockchain (Bitcoin, Ethereum, and Fabric) and studied in detail the applications that drove them (crypto, tokens, enterprise) and the area of blockchain technology that they pioneered (mining, smart contracts, and modularity).

In the chapters to come, we are going to look at how blockchain has become a nexus of other technologies and trends in the industry. The following chapter looks at blockchain's relationship to AI, IoT, and Quantum, and the chapters after that examine topics in Cybersecurity as well as blockchain's overall impact on the Web3 movement.

Chapter 7 –
ARTIFICIAL INTELLIGENCE, IoT AND QUANTUM

The thought of bringing AI and Blockchain together may be viewed by some as brewing a modern-day version of IT pixie dust. Let's let the magic begin.

COVERED IN THIS CHAPTER

- Blockchain brings trust to AI

- Blockchain and internet of things

- Blockchain and quantum computing

BLOCKCHAIN AT THE NEXUS OF TECHNOLOGY

Blockchain stands to accelerate the adoption of emerging technologies, including artificial intelligence, cloud, and internet of things by bringing in the missing element of trust, which is required by business to fully embrace these technologies at scale. On the flip side, blockchain business networks stand to benefit from the integration of these technologies into modern blockchain platforms and applications.

Given it's been a few chapters since I've told a story, I thought I'd start this chapter with the story of how blockchain is at the nexus of technologies like artificial intelligence, internet of things, and cloud computing.

THE STORY OF THE AI DOC AND AI OUTFITTERS

You picked your doctor because your doctor has a thing for technology, just like you do. And your doctor does not disappoint. She has a new AI Doctor's Assistant, which is expert trained, having ingested millions of medical periodicals to date. It would not be humanly possible for your doctor to keep up with all the latest medical breakthroughs given her busy schedule.

During my last checkup, I complained about some back pain. The doctor, with advice from the AI Doc, prescribed me Big Pharma's back pain relief capsules. The AI Doc also noted that my blood pressure has been trending slightly higher over the past year, and also advised my doctor to prescribe me Big Pharma's hypertension pills.

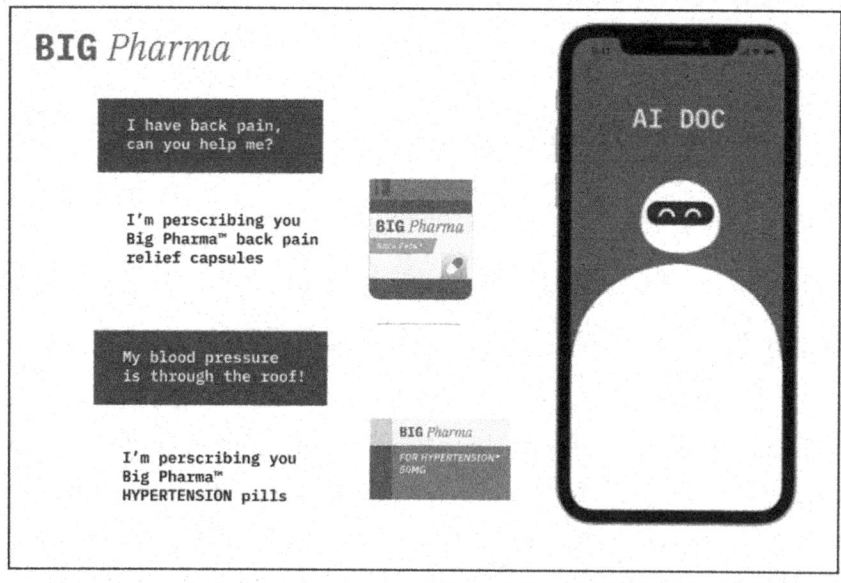

Chapter 7 – Figure 1 – Was AI Doc trained by Big Pharma? How would you know?

Okay. At this point, I start to wonder who provided the training data for the neural network that trained the AI Doc. Could it have been Big Pharma? If they were the sole providers of the training data, I would be skeptical about the objectivity of AI Doc. How can you (why should you) trust data from this or any other AI bot without knowing the provenance and authenticity of the training data?

As I head home, I check a mobile notification that says there is a 70% chance of rain today and offers me a coupon to buy an umbrella at AI Outfitters.

It never did rain that day, but two days later I received a second mobile notification from my weather app that there was an 80% chance of rain, and again, AI Outfitters offered me a coupon for a discount on a raincoat. It didn't rain again and I'm starting to think about where those weather sensors are located and who configured them. I cynically think maybe it's AI Outfitters and perhaps they tweaked the sensitivity of the weather sensors to increase the probability of reporting rain.

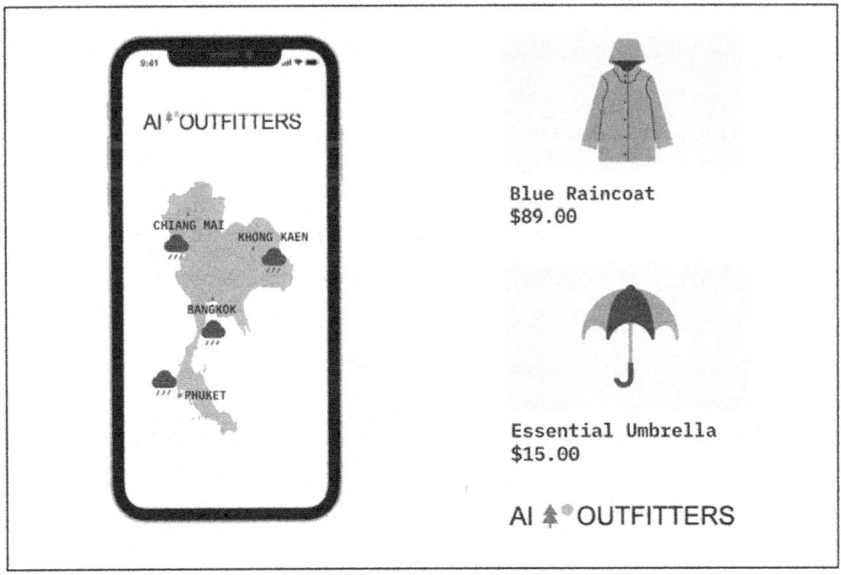

Chapter 7 - Figure 2 - Rain again? Can AI Outfitters ML models be trusted?

How can you (why should you) trust data from IoT sensors without knowing the provenance and authenticity of the sensors and that the data truly did come from the correct sensor, without tampered firmware, etc.?

Okay, if you are thinking these examples are a bit over construed, I wouldn't blame you. They are, but I am trying to make us think a little about the trust we have in our data today. Especially when you think about it from the AI, IoT and Cloud lens. How can we trust our AI-based recommendations, IoT sensor readings, or sensitive data uploaded to a public cloud? In this chapter, I touch on how blockchain stands to accelerate the adoption of emerging technologies including AI, Cloud, and IoT by bringing in the missing element of trust, which is required for businesses to fully embrace these technologies at scale. On the flip side, blockchain business networks stand to benefit from the integration of these technologies into modern blockchain platforms and applications. A win-win.

BLOCKCHAIN AND ARTIFICIAL INTELLIGENCE

Blockchain and AI are on just about every chief information officer's watch-list of game-changing technologies that stand to reshape industries. Both technologies come with immense benefits, but both also bring their own challenges for adoption. It is also fair to say that the hype surrounding these technologies individually may be unprecedented, so the thought of bringing these two ingredients together may be viewed by some as brewing a modern-day version of IT pixie dust. At the same time, there is a logical way to think about this mash-up that is both sensible and pragmatic.

Today, AI is for all intents and purposes a centralized process. An end user must have extreme faith in the central authority to produce a trusted business outcome. By decentralizing the three key elements of AI—that is, data, models, and analytics— blockchain can deliver the trust and confidence often needed for end users to fully adopt and rely on AI-based business processes.

Let's explore how blockchain is poised to enrich AI by bringing trust to data, models, and analytics.

Your Data Is Your Data

Many of the world's most notable AI technology services are centralized—including Amazon, Apple, Facebook, Google, and IBM, as well as Chinese companies Alibaba, Baidu, and Tencent. Yet all have encountered challenges in establishing trust among their eager but somewhat cautious users. How can a business provide assurance to its users that its AI has not overstepped its bounds?

Imagine if these AI services could produce a "forensic report," verified by a third party, to prove to you, beyond a reasonable doubt, how and when businesses are using your data once those are ingested. Imagine further that your data could be used only if you gave permission to do so.

A blockchain ledger can be used as a digital rights management system, allowing your data to be "licensed" to the AI provider under your terms, conditions, and duration. The ledger would act as an access management system storing the proofs and permission by which a business can access and use the user's data.

Trusted AI Models

Consider the example of using blockchain technology as a means of providing trusted data and provenance of training models for machine learning. In this case, we've created a fictitious system to answer the question of whether some fruit is an apple or orange.

This question-answering system that we build is called a "model," and this model is created via a process called "training." The goal of training is to create an accurate model that answers our questions correctly most of the time. Of course, to train a model, we need to collect data to train on—for this example, which could be the color of the fruit (as a wavelength of light) and the sugar content (as a percentage). With blockchain, you can track the provenance of the training data as well as see an audit trail of the evidence that led to the prediction of why a particular fruit is considered an apple versus an orange. A business can also prove that it is not "juicing up" its books by tagging fruit more often as apples, if that is the more expensive of the two fruits.

Explaining AI Decisions

The European Union has adopted a law requiring that any decision made by a machine be readily explainable, on penalty of fines that could cost companies billions of dollars. The EU General Data Protection Regulation (GDPR), which came into force in 2018, includes a right to obtain an explanation of decisions made by algorithms and a right to opt out of some algorithmic decisions altogether.

Massive amounts of data are being produced every second—more data than humans have the ability to assess and use as the basis for drawing conclusions. However, AI applications are capable of assessing large data sets and many variables, while learning about or connecting those variables relevant to its tasks and objectives. For this very reason, AI continues to be adopted in various industries and applications, and we are relying more and more on their outcomes. It is essential, however, that any decisions made by AI are still verified for accuracy by humans. Blockchain can help clarify the provenance, transparency, understanding, and explanations of those outcomes and decisions. If decisions and associated data points are recorded via transactions on a blockchain, the inherent attributes of blockchain will make auditing them much simpler. Blockchain is a key technology that brings trust to transactions in a network; therefore, infusing blockchain into AI decision-making processes could be the element needed to achieve the transparency necessary to fully trust the decisions and outcomes derived from AI.[46]

BLOCKCHAIN AND THE INTERNET OF THINGS

More than a billion intelligent, connected devices are already part of today's IoT. The expected proliferation of hundreds of billions more places us at the threshold of a transformation sweeping across the electronics industry and many other areas.

With the advancement in IoT, industries are now enabled to capture data, gain insight from the data, and make decisions based on the data. Therefore, there is a lot of "trust" in the information obtained. But the real truth of the matter is, do we really know where the data came from? And should we be making decisions and transacting based on data we cannot validate?

For example, did weather data really originate from a censor in the Atlantic Ocean? Or did the shipping container really not exceed the agreed tempera- ture limit? The IoT use cases are massive, but they all share the same issue with trust.

IoT with blockchain can bring real trust to captured data. The underlying idea is to give devices, at the time of their creation, an identity that can be validated and verified throughout their lifecycle with blockchain. There is great potential for IoT systems in blockchain technology capabilities that rely on device identity protocols and reputation systems. With a device iden- tity protocol, each device can have its own blockchain public key and send encrypted challenge and response messages to other devices, thereby ensur- ing a device remains in control of its identity. In addition, a device with an identity can develop a reputation or history that is tracked by a blockchain.[47]

Smart contracts represent the business logic of a blockchain network. When a transaction is proposed, these smart contracts are autonomously executed within the guidelines set by the network. In IoT networks, smart contracts can play a pivotal role by providing automated coordination and authoriza- tion for transactions and interactions. The original idea behind IoT was to surface data and gain actionable insight at the right time. For example, smart homes are a thing of the present and most everything can be connected. In fact, with IoT, when something goes wrong, these IoT devices can even take action—for example, automatically ordering a new part for your smart refrigerator. We need a way to govern the actions taken by these devices, and smart contracts are a great way to do so.[48]

In an ongoing experiment in Brooklyn, New York, a community[49] is using a blockchain[50] to record the production of solar energy and enable the purchase of excess renewable energy credits. The device itself has an identity and builds a reputation through its history of records and exchange. Through the blockchain, people can aggregate their purchasing power more easily, share the burden of maintenance, and trust that devices are recording actual solar production.

As IoT continues to evolve and its adoption continues to grow, the ability to autonomously manage devices and actions taken by devices will be essential. Blockchain and smart contracts are positioned well to integrate those capabilities into IoT.

BLOCKCHAIN & QUANTUM COMPUTING

Blockchain is revolutionizing transactions and business networks through trust, transparency, and security. Quantum computing will revolutionize computational power unlike the digital age has ever seen. But what do the two have in common?

Blockchain is often touted as a technology that is tamper-proof, or at the very least, tamper resistant. Primarily, this sentiment is due to standard cryptographic functions and consensus protocols that guarantee the security of a blockchain. These are relatively secure because breaking them requires huge computing resources, which are not generally available today. And yet blockchains may have an Achilles' heel.

Enter quantum computing. The highly touted security of blockchain is beginning to look like it is set to change with the emergence of powerful quantum computers. It will be child's play for such devices to break the kinds of cryptographic protection implemented in existing blockchain frameworks. Essentially, quantum computers could break the cryptography that conventional blockchains rely on, like the SHA-256 hash function that we studied in previous chapters.

While quantum computers can disrupt blockchain as it exists now, quantum cybersecurity can also provide a solution. Idalia Friedson, quantum computing expert and co-founder of the Hudson Quantum Initiative, suggests that incorporating emerging quantum cybersecurity in three steps can "save blockchain from the fate of other systems made obsolete by new technologies."

Step 1 - involves strengthening existing encryption algorithms by adding in truly random numbers, or so-called **quantum keys.** Adding quantum keys to blockchain software will provide added security against both a classical computer and a quantum computer.

Step 2 - involves developing **quantum-resistant algorithms**. The US - National Institute of Standards and Technology is currently reviewing submissions for these next-generation algorithms. One example is called Lattice cryptography that mathematically has been proven to be resistant to quantum computing attacks. So far, no known algorithms can break this method of encoding data.

Step 3 – involves using quantum key distribution hardware to send information from one point to another by encoding data on individual particles. Any attempted hack automatically severs the connection.

The threat posed by quantum computing to blockchain can be serious. But there are answers like Friedson's three-step plan to develop and implement quantum keys, quantum resistant algorithms, and quantum key hardware.

So, with a little forward thinking today, blockchain will continue to provide the foundation of trust through a tamper-resistant ledger and turn around the threat of quantum computing to actually enhance security levels.

- - -

As you can see, blockchain stands to accelerate the adoption of emerging technologies including AI and IoT by bringing in the missing element of trust. Similarly, blockchain business networks stand to benefit from the integration of these technologies into modern blockchain platforms and applications.

We have only just begun to witness blockchain's transformative power on business. As we look ahead to the future, we will almost certainly see blockchain effects multiply.

QUIZ TIME

Wednesday, April 13th, 2022 – Students. There is no homework today. The teacher is under the weather with a case of COVID-19. Other than what I would consider normal cold symptoms, I am doing fine, thank you very much. Hope you and your loved ones are staying well. I also hope you are enjoying this book… which will resume where we left off now.

UP NEXT

Blockchain is at the nexus of technologies like IoT, AI, and Cloud. It has the means of bringing the missing element of trust that is currently lacking from these technologies. Trust is gained through diversity of users. Unlike a database that has a single administrator, a blockchain enables a set of diverse administrators to "referee" the data, such that no single admin can maliciously or accidentally change or delete data. And once the administrators reach consensus, the data is secured in blocks that are "chained" to one another cryptographically forming a tamper-resistant ledger. When AI, IoT, and Cloud use blockchain to track provenance, proofs and permissions associated with data used and emanating from these systems, the trust in the data is increased radically. This trust will allow IoT, AI, and Cloud to be adopted without fear of compromise, ushering in a new era of applications and adoption of these technologies to change everyday life for the better. In data we trust… well, after you add a little blockchain.

- - -

The next chapter continues the trend of looking at blockchain at the nexus of technology. In this round, we will look at the critically important topic of cybersecurity. Specifically, we study four use cases that take advantage of blockchain's decentralized nature to ward off cyberattacks.

Chapter 8 -
CYBERSECURITY, ZERO-KNOWLEDGE PROOFS, DIGITAL IDENTITY

Blockchains possess ingredients that make it more naturally resistant to cyberattacks. Its inherently decentralized nature makes it a natural deterrent and well suited for cybersecurity use cases.

COVERED IN THIS CHAPTER

- How blockchain fends off the cyber threat

- Decentralized Cloud

- Zero-Knowledge Proofs

- Digital Self-Sovereign Identity

- Securing Software Supply Chains

THE CYBER THREAT

As the world becomes more digital, it has become equally vulnerable. Cyber-crime-related damage is projected to cost $6 trillion annually in 2021. Cyber-attacks are commonplace and have adversely demonstrated how they can disrupt our everyday lives. Governments worldwide saw a 1,885% increase in ransomware attacks, and the health care industry faced a 755% increase in those attacks in 2021, according to the *2022 Cyber Threat Report*.[51] These ransomware attacks hit supply chains, causing widespread system downtime, economic loss, and reputational damage, according to the report.

Looking at the related cybersecurity statistics and trends in the figure below, one can see data breaches and malware standout at the top offenders.

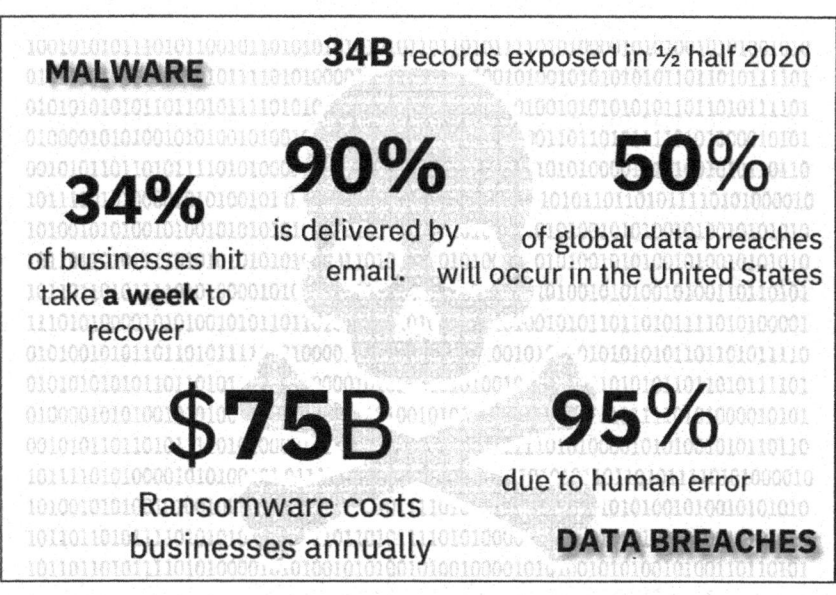

Chapter 8 – Figure 1 – Data Breaches and Malware top 2022 Cyber Threat Report

Blockchains possess ingredients that make it more naturally resistant to cyberattacks. The inherently **decentralized** nature of blockchain technology makes it a deterrent and as we will study in this chapter well suited for cybersecurity use cases.

Data is **immutable** on blockchains, meaning it is virtually impossible to be tampered with, as network nodes automatically cross-reference each other and pinpoint the node with misrepresented information. Blockchains also provide the highest standards of data transparency and integrity. As blockchain technology automates data management, it eliminates the leading cause of data breaches—human error.

Cybercrime is the greatest threat to enterprises and blockchain could go a long way in fighting it. What follows in this chapter is an overview of four use cases, where blockchain can be brought forward as a superhero whose superpowers come in the form of decentralization, immutability, and human error-resistant automation. A quick summary is provided, followed by a more complete discussion on each important topic.

Decentralized Cloud	Explores the vulnerability of modern hyperscale cloud services and how blockchain can be used to implement a new class of decentralized IT services to thwart centralized attacks.
Always Encrypted	Data breaches are amongst the top root causes of cyberattacks. Imagine a world where your data, once encrypted is never decrypted. If leaked, the criminals have gobbledygook.
Digital (Self-sovereign) Identity	Digital identity done right is a holy grail of cybersecurity. Using modern encryption and blockchain technology can keep control of personal identity information in the hands of its rightful owner... I.e., you.

Securing Software Supply Chains	Malware sneaks into software products like a bad virus. Paying attention to the provenance of our software, including where it came from, who wrote it, quality metrics, etc. Might do for software what food labels have done for food safety.

DECENTRALIZED CLOUD

In this section, we examine how blockchain-based cloud infrastructure can help alter cyberattack surfaces using decentralization. We focus on one specific attack (denial of service) and examine how blockchain services (decentralized domain name service and storage services) can effect positive change by defending our cloud computing platforms.

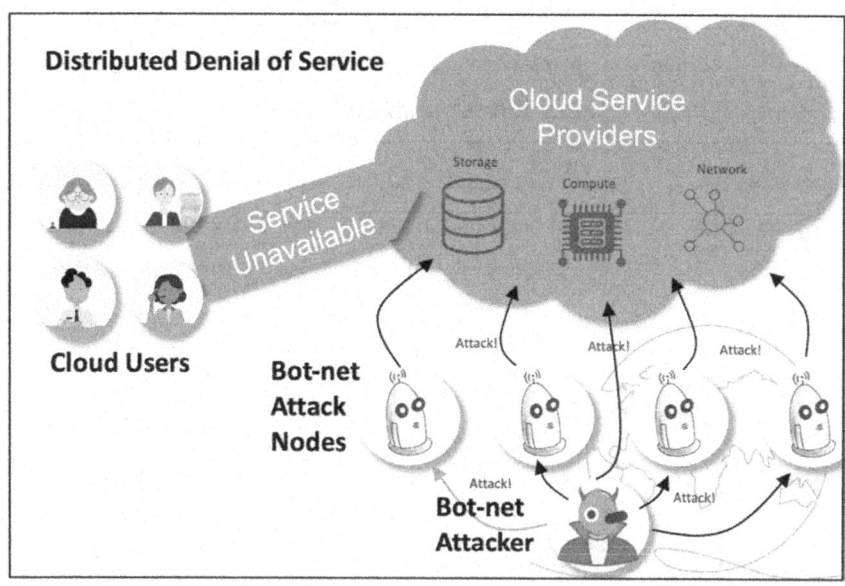

Chapter 8 – Figure 2 – Distributed Denial of Service Attack of Cloud via a Botnet

With more enterprises taking to the cloud, hyperscale cloud vendors such as Amazon and Google have been prime targets for cyberattacks. Many of such attacks take advantage of the decentralized nature of the Internet to both maintain anonymity and confuse security software. A common denial of service attack works by first infecting multiple nodes over a variety of domains to form a semi-coordinated network called a "botnet." These individual bots are then hijacked to launch attacks against targets that are much more centralized like today's clouds, often giving hackers an asymmetric advantage.[52]

Amazon	February 2020	A Distributed Denial-of-Service (DDoS) attack on Amazon Web Services (AWS) was the largest executed against AWS—and one of the largest publicly-disclosed attacks on any cloud up to that time.
Google	September 2017	Cyberattack launched against Google servers. The (DDoS) attack of foreign origin ramped up over a six-month campaign—and was the largest attack of its kind on record.

With blockchain's decentralized software deployment (smart contracts), data management (consensus-based transactions) and security protocols (encryption) could render targets less vulnerable by spreading attack surfaces and relying less on centralized trust.

At its peak, the 2017 Google attack was measured at 2.5 Tbps (terabits-per-second, a metric for comparing DDoS incidents), beating the previous record fourfold. Despite this unprecedented volume, Google was ultimately able to overcome the attack. DDoS attacks are designed to overwhelm a service with false requests from multiple maliciously infected sources. This ultimately cripples or completely shuts down the traffic on a targeted network or service. The decentralized nature of these attacks makes them difficult to pinpoint because there's no single point of origin to block.

While the attackers are decentralized, the targets of cyberattacks are largely more centralized. Servers often reside behind a single or limited number of IP addresses, providing a concentrated attack surface. Compromised passwords or cryptographic credentials can expose entire databases of valuable information. Hackers can take control of, or restrict access to, a large number of resources all at once, holding them for ransom.

Blockchain however, can level the playing field against hackers. Modern cloud designs are moving away from the traditional model of centralized trust, which creates a single point of failure, and toward a more "zero-trust" approach, especially with regard to security protocols. Distributing trust through consensus to validate important elements like access, authentication and database transactions is a function that blockchain is perfectly suited for.

What follows are two interesting examples of emerging decentralized cloud services for network and storage management.

Decentralized Network DNS

Many DDoS attacks exploit internet domain name servers (DNS)—which map IP addresses to readable website names. By moving DNS to blockchain, resources can be spread to multiple nodes, making it infeasible for attackers to control the database.

We're in the early days of decentralized DNS, and there are several active projects with similar goals. Namecoin was one of the first. It was released in 2011 but hasn't seen widespread adoption. Handshake is an interesting new contender that bills itself as a "decentralized naming and certificate authority."[53]

Unstoppable Domains is likely my favorite new entry in the decentralized DNS arena. Rather than using domain name services (DNSs) like traditional websites, unstoppable domains use a blockchain-based solution called the crypto name service (CNS). First released in 2018, Unstoppable Domains has grown in popularity very quickly throughout the cryptocurrency community.

Originally, the most common reason why someone would want an unstoppable domain was because having one would allow you to send or receive a variety of different types of cryptocurrency using a *human-friendly address* rather than the typical string of numbers and letters. However, these domains are now becoming increasingly popular to host normal websites and mobile apps as well.

I recently gave Unstoppable Domain (UD) a try for myself. With a few clicks, I was able to acquire my unstoppable domain. I'm now the proud owner of:

thinkblockchain.nft

Using UD was a delightful experience that almost makes thwarting cyber attacks fun.

Filecoin - Decentralized Cloud Storage

In contrast to centralized, permissioned cloud providers, decentralized cloud storage providers leverage infrastructure that is designed to mitigate undue control or influence. These providers typically also utilize a permissionless structure that enables developers to employ their services with reduced restrictions. Conceptually similar to a decentralized blockchain, decentralized storage models draw their security from their widely distributed structure. This overall architecture can help make these systems more resistant to the hackers, attacks, and outages that have plagued large, centralized data centers.

Created by the same team responsible for the Interplanetary File System (IPFS), Filecoin is an open-source, cloud-based storage network that aims to improve upon some perceived shortcomings of traditional cloud storage providers. These pain points include insufficient trust, unreliable security, limited connectivity, poor scalability, and heightened dependency on centralized systems.

Filecoin's goal is to build a large, decentralized file storage network that can be adapted to meet the varying needs of different customers. Low costs, fast retrieval speed, and data storage redundancy are a few of the modifications that may appeal to customers who choose to use Filecoin. Filecoin also claims to be built with state-of-the-art cryptographic storage proofs to ensure data is being stored securely and precisely for a specified time-period.[54]

- - -

The essence of asymmetric warfare is demonstrated by the 2017 Google DDoS attacks, taking advantage of the fact that hackers are well-distributed with respect to their targets. By decentralizing assets, applications and security infrastructure using blockchain, it may be possible to stop fighting hackers on their terms and beat them at their own game.

ALWAYS ENCRYPTED

You enter a bar and are asked to present your government ID to prove you are of legal age (21 in North Carolina) to enter this establishment where alcohol is served. You pull out your State issued driver's license and hand it to the bar worker. The worker scans it, types a few things on his keyboard and gives you the go ahead to enter. As you pass, he says, "Go State!"

While I find that encounter amusing, it's worth reflecting on what happened. While the document that I shared had the information the bouncer needed, it had much more information than required to get the job of proving my age done. In fact, the worker clearly looked at my address as well, which led to the correct realization that given I live in Raleigh, I must be an NC State fan.

This section explores two advanced cryptographic techniques that might seem like magic. (I know they did to me when I first learned about them). These technologies would allow me to share an encrypted version of my license with the bar worker, and without ever decrypting it to unveil my address, eye color, height, or organ donor status, prove with the highest degree of certainty that I am indeed old enough to enter (and God knows, I am indeed old enough).

Homomorphic Encryption and Zero-Knowledge Proofs are two trending concepts that are widely popular as data privacy preservation techniques in a wide variety of applications, especially in those associated with blockchain technology which are immutable, distributed and secure.

What follows is a quick summary of each of these revolutionary technologies, followed by a deeper explanation of how they can be effectively used to improve the privacy of blockchain applications in various domains.

Homomorphic encryption	A cryptographic technique that allows you to perform computations on encrypted data without decrypting it.
Zero-Knowledge Proofs	A cryptographic technique providing proof that a certain statement is correct, without revealing any details about the statement.

Homomorphic Encryption

While the state of the art of today's Internet security liberally applies encryption to protect data (e.g., https), the sensitive data typically must first be decrypted to access it for computing and business-critical operations. This opens the door to potential compromise of privacy and confidentiality controls. Until now, those vulnerabilities have been the cost of doing business in the cloud and with third parties.

An innovative technology, Fully Homomorphic Encryption (FHE), is poised to achieve zero-trust by unlocking the value of your data on untrusted domains without needing to decrypt it.

The classic approach to data security is declining in effectiveness as we move towards a future that features growing privacy regulation and increased data sharing in uncontrolled environments. What if organizations were able to compute upon sensitive data while the data itself remained in an encrypted state? Several decades ago, cryptographers asked the same question—what if data could remain encrypted while it's computed upon, therefore preserving the confidentiality (and privacy) of the data as the computations are done? Theory became reality in 2009 when Craig Gentry showed FHE was plausible.

However, processing even a single bit took massive computing power and too much time. Over the past decade, researchers, such as those at IBM Research, have been working steadfastly to make FHE more practical and make it much more efficient. IBM Research developed open source libraries, such as the FHE Toolkit, which have brought this game-changing domain of cryptography to more people.

So why is FHE not in broad use today? Simply put, various hurdles still exist for organizations to adopt this technology. The industry is still in the early days of shifting FHE from research labs into commercial applications. While fully homomorphic encryption has become more efficient over the years, certain operations can require hundreds of times the computational resources when compared with an equivalent operation on plaintext data. That said, given the value and sensitivity of a given transaction, the additional compute cost might be a small price to pay for peace of mind.

Zero-Knowledge Proofs

While homomorphic encryption ensures privacy by allowing computations on encrypted data, making it more suitable for scenarios where analysis of confidential data without compromising privacy is sole priority, Zero-Knowledge Proofs (ZKP) are more suitable for generating proofs for personal attributes and transactions in applications that need to preserve the identity of participants.

Zero-Knowledge Proofs are an amazing and counterintuitive cryptographic concept, first proposed by Goldwasser, Micali, and Rackoff in a paper that introduced the idea of *interactive proof systems*. Like the "getting carded at the bar" example, ZKP provides privacy in situations where you'd otherwise have to reveal confidential information. Examples include:

Logging into websites	Instead of typing your password into a potentially unsafe website, you can simply send a proof that you "know your password"
Authenticating your identity	Instead of giving your mother's maiden name over the phone to a random bank call center agent, you can simply send a proof (a cryptographic fingerprint) that you are who you say you are
Sending private blockchain transactions	Instead of sending money on Bitcoin, where your financial ledger is public information, you can send a proof that your money is valid (not double spent) without revealing your balances, as popularized by ZCash

The following example is one of my all-time favorites and my rendition is inspired by a paper written by Naor, Naor, and Reingold, titled "Applied Kid Cryptography or How to Convince Your Children You Are Not Cheating".[55] In my version, Jane and Howard are racing to find Donald in a popular children's book series, where the point is to spot Donald in a sea of shapes that look like him.

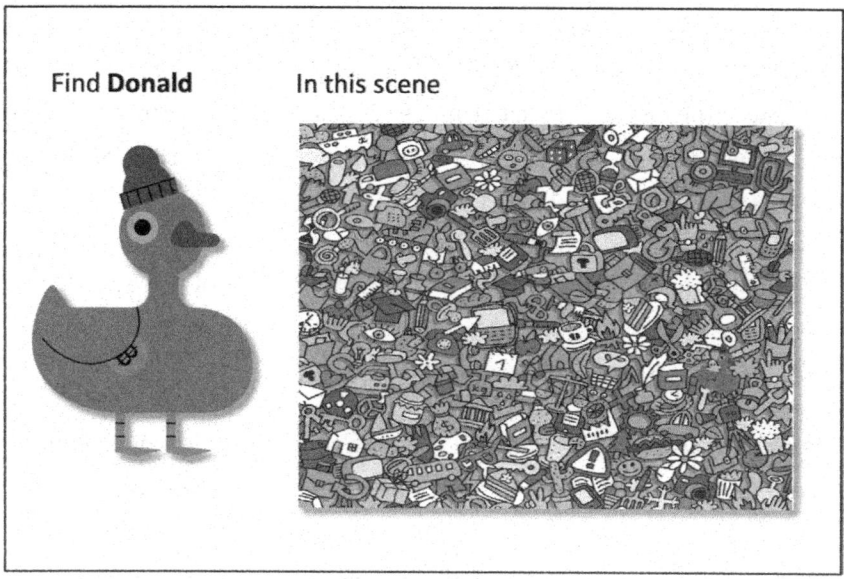

Chapter 8 – Figure 3 – How does Jane prove to Howard that she knows where Donald is?

The puzzle:

Jane: I know where Donald is!

Howard: Jane, have you ever heard the phrase "liar, liar pants on fire"?

Jane: I can prove to you where he is without revealing his location.

To defend her integrity, Jane devises two solutions to prove her knowledge.

Proof 1 – Jane cuts out Donald from her scene and only shows Howard the Donald snippet. To ensure that Jane didn't just print out a new picture of Donald, Howard can watermark the back of Jane's scene page.

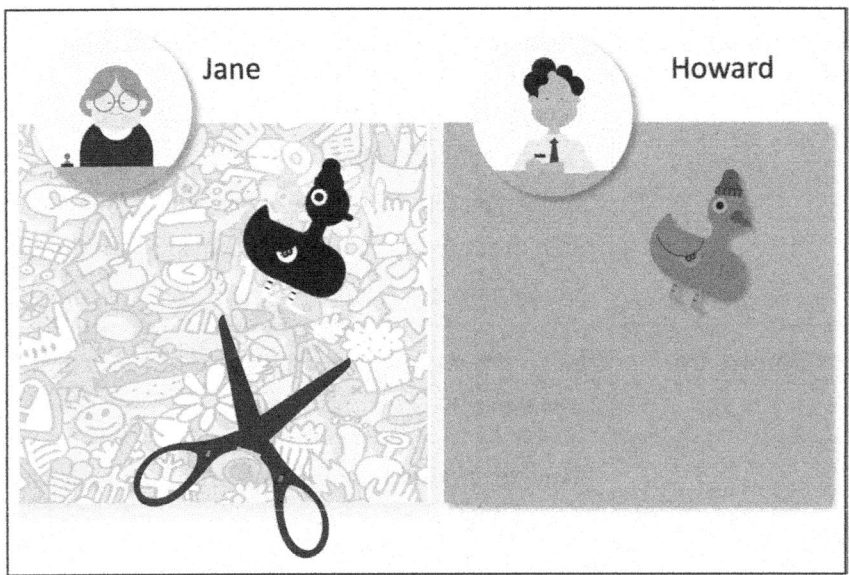

Chapter 8 – Figure 4 – Jane creates a ZKP using scissors and cardboard

Proof 2 – Jane cuts a hole in a very large, opaque sheet of cardboard. She places the cardboard cutout on top of the original scene. In this solution, only Donald is shown. His coordinates relative to the rest of the scene is still unknown. Later, Jane can reproduce the scene underneath to prove that she used the original puzzle.

Soundness, Completeness, and True Zero-Knowledge
This solution fulfills the three important properties of zero-knowledge proof systems: soundness, completeness, and true zero-knowledge.

Jane can use the same proofs to verify that she has found Donald many times per game, and across many games. In this sense, her proof systems achieve statistical:

Soundness	*Jane's proofs are truthful and do not let her cheat.*
	Everything that is provable is true: Assuming Jane doesn't know Donald's locations and presents random pieces of the scene to her proof systems… then, her cardboard holes will display random images without Donald.
Completeness	*Jane's proofs convince Howard that she found Donald.*
	Everything that is true has a proof: As long as Jane finds Donald, she's able to consistently use her proofs to show Donald, in each game.
True Zero-Knowledge	*Jane's proofs prove her victory to Howard, without revealing her knowledge.*
	Only the statement being proven is revealed: As Jane proves to Howard that she has found Donald, the only information revealed to Howard is that "Jane has found Donald." Donald's location is never revealed.

- - -

Logging into an online account without a password. Completing a cryptocurrency transaction without revealing your balance. Verifying Covid-19 status without specifying vaccination, previous infections, or test dates. Verifying household income for a loan without disclosing your income. The concept of enabling verification of a statement without disclosing the actual information is at the very core of Zero-Knowledge Proof (ZKP).

ZKP fills gaps left vulnerable by the anonymity provided by some distributed ledger technologies. ZKP raises the bar for certainty in cryptography because a true piece of zero knowledge is virtually impossible to break. "It transforms the basic commodity of trust from being synonymous with crossing your fingers and lack of privacy, into a real, unbreakable and 100% private, virtual, good".[56] As such, ZKP enables self-sovereign identities, which are covered in the next section.

DIGITAL (SELF-SOVEREIGN) IDENTITY

Imagine a world in which you always have peace of mind that your personal information is safe. Imagine a world in which your information cannot be shared without your clear, explicit consent at the time of the transaction; where you decide who can access what information, when, and for how long. In this world, you can even later choose to revoke that privilege. You are in control. Every person, organization, or thing can have its own truly independent digital identity that no other person, company, or government can take away. Digital Identity was one of the Three Blockchain Stories that we covered in Chapter 2. In this section, we go a little deeper into what it means to have self-sovereign identity and how this can be the single most important blockchain-based advancement to secure our cyberspace.

Today, we are not in control of our identity. Our personal information lives in centralized repositories outside of our control. Information is often shared without our awareness. Daily, we hear stories of security breaches and identity theft that erode our confidence and trust. The hope on the horizon is self-sovereign identities.

SSI

Self-Sovereign Identity (SSI) offers better security, privacy, and safety than digital identities like Facebook or Google-based sign-on. Digital identity usage is increasing among the public and businesses. You can use your Google or Facebook profile to sign on to public websites like Trello, Medium, eBay, Instagram, and so on. However, these digital identities are prone to hacking incidents and may reveal more information about you than you realize.

SSI is a rising trend in digital identity that functions as a tamper-resistant digital identity based on blockchain technology. When you use SSI to log in for any services, you'll control the data related to your identity and not the service provider. Online service providers will also experience increased system efficiency and business opportunities by giving you more control over your data.

How does SSI work?

SSI works in a simple way, where an issuer will issue digitally signed documents for you. The issuer will also establish document trust through a blockchain network. You'll receive your documents in a digital wallet, most likely a blockchain-based app. You could even use government and private services, using the SSI wallet app to authenticate your identity digitally.

There are three key concepts that make SSIs work, which are explored in the sections that follow.

1. Verifiable credentials

2. Decentralized Identifiers

3. Blockchain

VCs

Verifiable Credentials (VCs) are an open standard for digital credentials. They can represent information found in physical credentials, such as a passport or license, as well as new things that have no physical equivalent, such as ownership of a bank account. They have numerous advantages over physical credentials, most notably that they're digitally signed, which makes them tamper-resistant and instantaneously verifiable. Verifiable Credentials can be issued by anyone, about anything, and can be presented to and verified by everyone. The following figure describes the personas and roles involved to complete an SSI exchange.

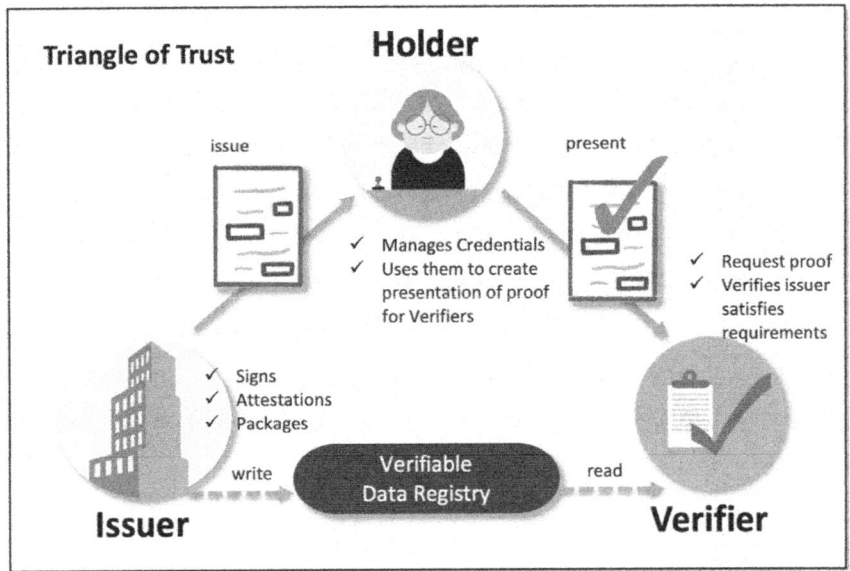

Chapter 8 – Figure 5 – Verifiable credential sits at the center of a triangle of trust

The entity that generates the credential is called the *Issuer*. The credential is then given to the *Holder,* who stores it for later use. The Holder can then prove something about themselves by presenting their credentials to a *Verifier.*

The holder of a Verifiable Credential sits at the center of a triangle of trust,[57] mediating between issuer and verifier.

- The issuer trusts the holder

- The holder trusts the verifier

- The verifier trusts the issuer

Any role in the triangle can be played by a person, an institution, or an IoT device.

Note that because verifiable credentials can be created by anyone, the person or entity verifying the credential decides if they trust the entity that issued it. It is like a shop clerk deciding if they should accept an out-of-state license as proof of age when purchasing alcohol.

The VC model places the holder of a credential at the center of the identity ecosystem, giving individuals control of their identity attributes. The W3C VC model parallels physical credentials: the user holds cards and can present them to anyone at any time without informing or requiring the permission of the card issuer. Such a model is decentralized and gives much more autonomy and privacy to the participants. This contrasts with the federated identity management (FIM) model, as adopted by SAML and OpenID Connect, which place the identity provider (IdP) in the central role as the dispenser of identity attributes and the determiner of which Service Providers (SPs) it will give them to. In the federated model, the IdP knows every SP that the user visits.

I found it helpful to actually see how a credential is represented in JSON. What follows is a verifiable credential using the W3C VC definition. This specific credential is a Duck Hunting License issued by the State of North Carolina. However, you can easily imagine a similar credential that proves you have a Bachelor's Degree from Acme University.[58]

```
 1   "verifiableCredential": {
 2       "@context": [
 3           "https://www.w3.org/2018/credentials/v1",
 4           "https://www.w3.org/2018/credentials/examples/v1"
 5       ],
 6       "id": "0892f680-6aeb-11eb-9bcf-f10d8993fde7",
 7       "type": [
 8           "VerifiableCredential",
 9           "Fishing Hunting License"
10       ],
11       "issuer": {
12           "id": "did:example:76e12ec712ebc6f1c221ebfeb1f",
13           "name": "North Carolina Game and Wildlife"
14       },
15       "issuanceDate": "2021-05-11T23:09:06.803Z",
16       "credentialSubject": {
17           "id": "did:example:ebfeb1f712ebc6f1c276e12ec21",
18           "degree": {
19               "type": "DuckHunting",
20               "name": "Duck Hunting"
21           }
22       },
23       "proof": {
24           "type": "Ed25519Signature2018",
25           "created": "2022-04-19T11:33:12Z",
26           "jws": "eyJhbGciOiJFZERTQYjY0Il19..nlcAA",
27           "proofPurpose": "assertionMethod",
28           "verificationMethod": "https://pathToIssuerPublicKey"
29       }
30   }
```

DIDs

Decentralized Identifiers (DID) are what enable Verifiable Credentials to be verified anywhere, at any time, even if the issuer does not exist anymore.

In the example above, lines 12 and 17 are DIDs. DIDs can be thought of as being similar to a URL (Uniform Resource Locator) aka a web address. It's an identifier for something, only in the case of a DID, this identifier is based on decentralized technology (by contrast URLs are managed through domain registries, which are usually centralized).

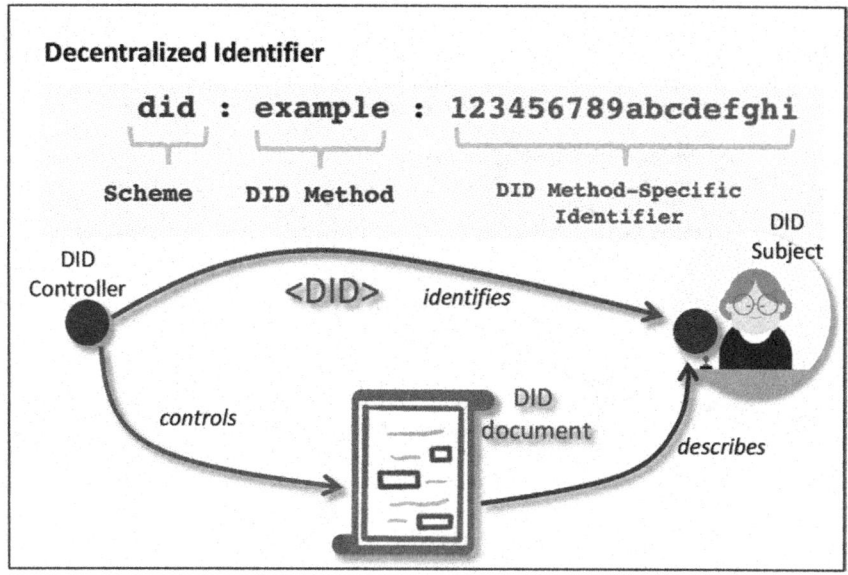

Chapter 8 – Figure 6 – URLs are to websites as DIDs are to digital identity

The DID draft published by w3c, declares a Decentralized Identifier as a simple text string consisting of three parts:

- The DID URI scheme identifier, stored on-chain

- The DID method identifier

- The DID method-specific identifier

DIDs are part of a global key-value database wherein compatible blockchains such as Ethereum and Sovrin host the DID Documents (e.g., public keys, service endpoints and authentication protocols). DIDs act as keys and DID Documents as values to describe specific data models to bootstrap cryptographically verifiable interactions with the identified entity in the decentralized ecosystem.

Blockchain

In a scenario where blockchain is used for identity management, no Personally Identifiable Information (PII) is stored on the blockchain. This is crucial as a distributed ledger is immutable, meaning anything that is put on the ledger can never be altered nor deleted, and thus no personal data should ever be put on the ledger. Only the issuer's Public DID is stored on the ledger.

No PII on ledger, then why blockchain? So, what problem is blockchain solving for identity, if PII is not being stored on the ledger? The short answer is that blockchain provides a transparent, immutable, reliable, and auditable way to address the seamless and secure exchange of cryptographic keys. Hence, blockchain is used for managing crypto-keys, proof, and permissions only. All PII is kept off-chain—for example, in a digital wallet on your smart phone.

Hyperledger Indy provides tools, libraries, and reusable components for providing digital identities rooted on blockchains or other distributed ledgers so that they are interoperable across administrative domains, applications, and any other silo. Indy is interoperable with other blockchains or can be used standalone to power the decentralization of identity.

SSI for COVID Verification

SSI technology benefits both end-users and service providers alike. New systems based on SSI are demonstrating instant value because they are transparent, easily auditable, free from human error, and more secure. Here is an example that is near and dear.

IBM and the state of New York have brought Excelsior Pass, a New York State branded Health Pass, to the market. The **first State Health Pass** to be rolled out in the United States.

The Excelsior Pass was launched on March 25th, 2021, making the tool available to all 19 million New Yorkers and all businesses across the state. The application has three components—the portal, wallet, and scanner. Those who have been vaccinated or tested can visit the portal—a website for users to receive their Excelsior Pass. After they have successfully received their pass, they can download the NYS Excelsior Pass Wallet application to their smartphone as a place to store their valid credential. The credential is what people will show to businesses accepting Excelsior Pass and looks quite similar to an airline boarding pass. The third and final component is the scanner. Businesses can download the NYS Excelsior Pass Scanner to a smartphone or tablet and scan the QR codes of patrons' passes to verify they have been tested or vaccinated.

The Excelsior Pass Plus is powered by IBM Digital Credentials, a blockchain-based platform that uses W3C verifiable credential standards to provide individuals and organizations with the core capabilities they need to securely issue, manage, and verify digital credentials. Proof of vaccination or a negative test result is auditable, traceable and verifiable—in seconds.

- - -

Today, most risks facing consumers from cyber criminals can be tied back to credential or data theft in some form. Self-Sovereign Identity standards are key elements of a future that provides increased security and risk protection for identities, data, and online interactions. Subsequently, one of today's biggest risks facing consumers from cyber criminals will be successfully mitigated.

Next, we will examine the fourth and final topic of cybersecurity, which is securing the software supply chain.

SECURING SOFTWARE SUPPLY CHAINS

In the context of the modern-day approach to software development, I often joke by saying that the last unique line of code was written in 1997. From there, the process of development comprised the following tasks: Google-search for a code snippet, cut-and-paste, debug, and repeat until done. Recent research supports this, showing that most of the software produced today is not written from scratch, but rather cobbled together from existing software components or libraries. Application developers are increasingly reliant on open source component parts because pre-fabricated components speed up innovation and save developers the time (and money) of having to write code from scratch.

But with 6.1% of component downloads containing a known security vulnerability, it's inevitable that defective or malicious parts will make their way into production—especially with component management practices lagging. Up until recently, it's been difficult for organizations to fully grasp the enormity of what it means to have to work backwards to fix the use of defective, outdated, and risky components in applications.[59] The application security community has long recognized the risk involved in using vulnerable components. In 2013, the famous OWASP Top Ten introduced "A9 Using Components with Known Vulnerabilities". A9 recommends organizations "continuously inventory the versions of both client-side and server-side components (e.g., frameworks, libraries)". A major inhibitor to cyberattacks is understanding the components in an application, especially which of those are vulnerable.

In February of 2022, The Linux Foundation, of Hyperledger-fame, covered in a previous chapter, announced the availability of the first in a series of research projects to understand the challenges and opportunities for securing software supply chains. "The State of Software Bill of Materials and Cybersecurity Readiness" reports on the extent of organizational SBOM readiness and adoption tied to cybersecurity efforts.

SBOM

An SBOM or Software Bill of Materials is an itemized list of components that make up a software application. Bills of materials have been in use in the manufacturing industry for decades in order to track the required inventory for production and ensure replicability in the outputs. For those familiar with

DevOps, this manufacturing analogy may sound familiar. In the same way, manufacturers judiciously manage the parts for building products, SBOMs help developers track their dependencies in a more shareable format and align third party library usage across application teams and reduce wasteful "component sprawl" that comes from the "Google-search/cut-and-paste" method described above.

The Linux Foundation study comes on the heels of both the U.S. Administration's Executive Order on Improving the Nation's Cybersecurity and the recent White House Open Source Security Summit. Its timing coincides with increasing recognition across the globe of the importance of identifying software components and helping accelerate response to newly discovered software vulnerabilities.

"SBOMs are no longer optional. Our Linux Foundation Research team revealed 78% of organizations expect to produce or consume SBOMs in 2022," said Jim Zemlin, executive director at the Linux Foundation. "Businesses accelerating SBOM adoption following the publication of the new ISO standard (5962) or the White House Executive Order, are not only improving the quality of their software, but they are also better preparing themselves to thwart adversarial attacks following new open source vulnerability disclosures like those tied to log4j".[60]

More specifically, an SBOM is formal and machine-readable metadata that uniquely identifies a software component and its contents; it may also include copyright and license data. SBOMs are designed to be shared across organizations and are particularly helpful at providing transparency of components delivered by participants in a software supply chain. Many organizations concerned about application security are making SBOMs a cornerstone of their cybersecurity strategy.

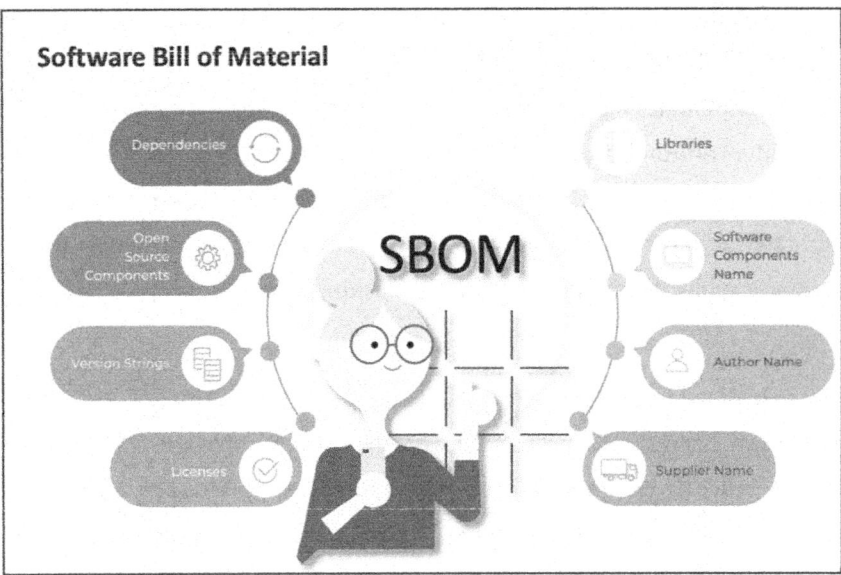

Chapter 8 – Figure 7 – Software Bill of Material definition and lifecycle

An SBOM contains the following information.

Data fields	As shown in the figure above, including supplier name, component name, version of the component, etc.
Operational considerations	Frequency of SBOM generation, depth of the dependency tree, and access to SBOM data
Support for automation	Ensures the data can be produced at scale and consumed at scale using three different data formats already standardized, including three leading file formats known as SPDX, CycloneDX, and SWID

This past year, we have seen several unprecedented attacks rooted in our software supply chain, including the Microsoft Exchange server attacks, the Kaseya ransomware attack, and the SolarWinds breach. Our traditional methods of cybersecurity are failing, and SBOMs are only one piece of the prevention puzzle.[61] "The SBOM is a necessary, not a sufficient remedy for the large, global, systemic challenges of software development. Other things are needed—a global regulatory framework that forces the market into action, better technologies—e.g. a blockchain system for the software supply chain—and better training programs that will raise awareness," said Dr. Rispens in a well-written blog on the topic titled, "Why the World Needs a Software Bill Of Materials Now".[62] The following section expands on Dr. Rispens' call-to-action on a blockchain-based software tracking system.

Decentralized SBOM Registry

There have been several proposals to create a decentralized SBOM Registry using blockchain technology. The gist of these solutions is to apply blockchain technology as a SBOM ledger, where the data of each software dependency can be recorded, hashed, and stored. Each change or addition to the software creates a new block in the chain without altering existing records. If an actor tries to alter the existing record, the change will appear as an altered hash value for all subsequent blocks in the chain.

For example, if an original software package had a component acquired from Huawei Technologies, the SBOM recorded through blockchain would include that information. If the supplier attempted to remove the data indicating the provenance of that software component, the blockchain would detect a change to the hash value. The consumer could then see that the SBOM received does not match the publicly available ledger, meaning there is a problem.

The use of blockchain as an integrity solution could be further strengthened by pairing it with a restoration process for **planning, monitoring, and tracking** software updates.[63]

A CYBERSECURITY CALL-TO-ACTION

While this chapter called out four topics where blockchain can be used to improve our situation in cybersecurity, one can imagine several more. I came across another well-written blog titled, "The Future Use Cases of Blockchain for Cybersecurity", where Julien Legrand covers a few use cases not covered in this chapter, including:[64]

Securing Private Messaging	Most messaging companies are warming up to blockchain for securing user data as a superior option to the end-to-end encryption, which they currently use.
IoT Security	Blockchain can be used to secure such overall systems or devices by decentralizing their administration. The approach will give the capabilities of the device to make security decisions on their own.
Verification of Cyber-Physical Infrastructures	Data tampering and systems misconfiguration together with component failure have marred the integrity of information generated from cyber-physical systems. However, the capabilities of blockchain technology in information integrity and verification may be utilized to authenticate the status of any cyber-physical infrastructures.

Perhaps this chapter is the most critical of the entire *Think Blockchain* book. In that the thesis of the book is blockchain's applicability to use cases across all industries. However, the growing threats of cyber-attacks stunt the growth of emerging technologies and the markets they support. Therefore, the call-to-action is to dig deeper into these subjects introduced in this chapter and find a way to do your share to utilize them.

Remember that no matter how it is utilized, the key component of blockchain technology is its ability to decentralize. This feature removes the single target point that can be compromised. As a result, it becomes practically impossible to infiltrate systems or sites whose access control, user-identities, data storage, and network traffic are no longer in a single location. Therefore, blockchain may be one of the most efficient mitigation strategies for cyber threats in the coming days. Nevertheless, blockchain, just as with any other new technologies, faces many startup challenges as it undergoes the process of growth. But like the old New York State Lottery slogan states, "You've got to be in it, to win it." There is no better time than now to take a run at cyber threats with blockchain technology.

QUIZ TIME

1 – What is the unfair advantage that botnet leverages in a DDoS attack on a cloud?

    a.   The attackers and targets are both decentralized

    b.   While the attackers are decentralized, the targets of cyberattacks are largely more centralized.

    c.   The attackers use email phishing as a means to coordinate bots using malware within the compromised email accounts

    d.   None of the above

2 – Decentralized Identifiers (DIDs) can be thought of as being similar to?

    a.   A digital wallet that contains your credit cards, licenses, and certification cards

    b.   A driver's license or fishing license

    c.   A Uniform Resource Locator (URL) that uniquely identifies a website

    d.   All of the above

3 – Zero-Knowledge Proof systems must achieve the following to be statistically sound.

   a. Perform operates on encrypt data using homomorphic encryption

   b. Give 100% guarantee that the claim is valid

   c. Achieve soundness, completeness, and true zero knowledge

   d. Must be Turing Complete

4 – A Software Bill of Material (SBOM) is formal and machine-readable metadata that uniquely identifies a software component and its contents and can help avoid security vulnerability because your software product is known and accounted for?

   a. True

   b. False

5 – Think Blockchain Labs– Chapter 8 Assignment

Go to the Think Blockchain Labs Github repository.

See the README.md file and search for the Chapter 8 assignment details, but the gist of the assignment is to give:

   a. A coding example that further explores Zero-Knowledge Proof

   b. A coding example that further explores Self-Sovereign Identity

UP NEXT

I hoped this chapter wasn't like drinking water from a firehose. We certainly covered a great deal of ground on the topic of securing our growing digital economy. This topic is both extremely relevant and timely. And as you can tell from my tone and comments made during this chapter, cybersecurity is one topic that I'm passionate about. I am fortunate to have opportunities to work with the United States Government on topics related to improving our nation's cybersecurity posture. In fact, digging through my camera roll, I was only able to find this photo of the 2016 Cybersecurity panel that President Obama commissioned. You might also enjoy reading the testimony that I delivered to this commission. It can be found in the Addendum 1 – Chapter of this book, which contains my government-focused testimonies. Fun fact: Irving Wladawsky-Berger, who authored the foreword of this very book, was also an expert witness at this 2016 commission hearing. Irving has a wonderful way of filling a room with energy, as he did that day.

Chapter 8 – Figure 8 – Jerry's view from the "hot seat" at
the 2016 National Cybersecurity Commission

– – –

We're down to one more technology-oriented chapter. It's fitting that this chapter anchors the content covered thus far because Web3 can very well be the phenomenon that brings all the previous concepts of this book together in a more unified and sensible way. Many say Web3 is the future of the internet. Skeptics say, it's just a buzzword. However, there is no doubt in my mind that the wonderful thing that we call internet will evolve. The next wave must tackle the toughest challenges covered within this Cybersecurity chapter, while also providing a means to flow more of the control and rewards to those creating the value on the web. What follows is one author's view of what Web3 is and/or could be in the decades to come.

Chapter 9 -
A ROAD TO WEB3

Is it the Internet of the future? A more decentralized internet? A "stateful internet" of value or is it just a buzzword?

COVERED IN THIS CHAPTER

- Web3-like shifts seen before

- How Web3 builds on Web2

- Web3 Architecture and developer experience

- Airing Web3 grievances and solutions

THE RISE OF THE CREATORS

Welcome to the late 1880s. The relatively new National Baseball League had been around for a little more than a decade—enough time for labor relations to have grown tense. The players felt that the owners' vision for the game was fundamentally untenable for them. In short, they were the stars, yet the owners made the lion's share of the money. After all, it was the players that hit the home runs that put fans in the stands. So, that year, the players set off on their own. They had an idea for a version of baseball, where they would have both a share of the profits and a say in the operations.

The Players' League was born. The name captured the idea—a league by and for the players who made it. They were the creators and the source of the value and entertainment that fans paid for. The Players' League was their radical concept that not only revolutionized organized labor in sports, but for organized labor, period. They went head-to-head with their old bosses in the National League, and initially, they were successful and drew more fans.

The Players' League only lasted a year. However, it remains among the most interesting experiments in sports labor history, and its cause still resonates more than a century later. "The issues that animated labor for baseball players in the 1880s have not disappeared," says MLB's official baseball historian, John Thorn. "A lot of that's going to sound familiar."[65]

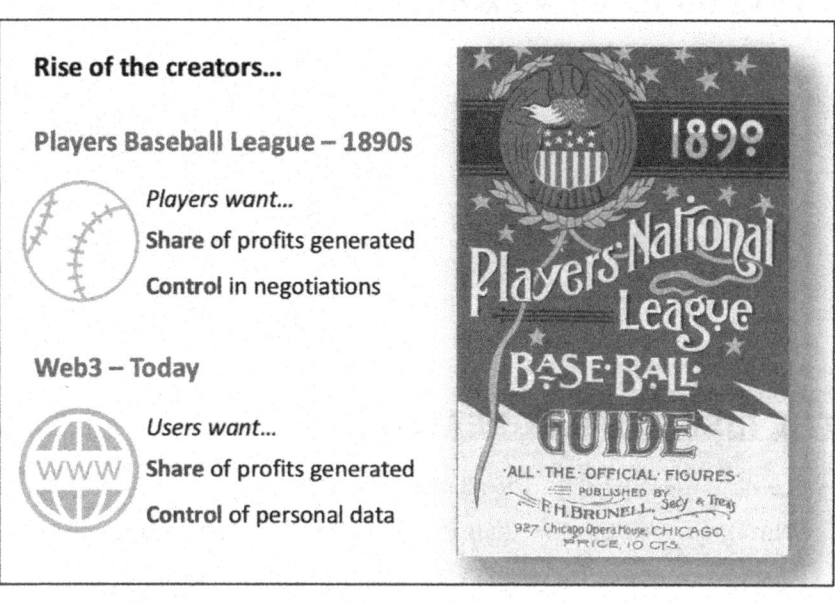

Chapter 9 – Figure 1 – Players League aimed to shift control and profit to creators

The Players' League story is a rebellion seeking balance that favors the interests of "the creators." There are many other examples spanning widely across the media and entertainment industry. The NFL, NBA and Premier (Soccer)

League (just to name a few) have all seen radical labor shifts towards the players. Music rights have shifted similarly, from the big music intermediaries like Sony and RCA to individuals self-publishing their music on Bandcamp.com.

This ideal of shifting **profit and control to the creators** is the general movement behind Web3. So, in a sense, we have seen this behavior before. But instead of the centralized, owners-led entertainment examples outlined above, today it's the dominance of big tech over internet use and its control over personal data that has led calls for a third wave of the internet.

EVERYWHERE I GO "WEB3, WEB3, WEB3". WHAT IS IT?

The term "Web3" was coined in 2014 by Ethereum co-founder Gavin Wood, to capture an idea of a new kind of internet service that is built using decentralized blockchains. Since then it's become a catch-all term for anything that has to do with the next generation of the internet being a decentralized digital infrastructure. The movement gained interest in 2021 from cryptocurrency enthusiasts, large technology companies, and venture capital firms.[66] Packy McCormick, an investor who helped popularize Web3, has defined it as "the internet owned by the builders and users, orchestrated with tokens".[67]

The current internet, Web 2.0 (Web2), relies on systems and servers owned largely by big corporations, raising concerns over system vulnerability and control. Advocates of Web3 are proponents of internet activity being governed by "the many" rather than the incentives and biases of the few, with the biggest concerns being large corporations controlling and disproportionately profiting from personal data.

The theory is that in a Web3 world, sensitive activities and data would be hosted on a network of computers using blockchain rather than corporate servers. The internet would likely have the same look and feel, at least initially, but your internet activities would be represented by your crypto-wallet and websites hosted through decentralized applications (DApps) run on a blockchain network. The following table provides a compare-and-contrast summary of Web1, Web2 and Web3.

Web1	Web 1.0 refers roughly to the period from 1991 to 2004, where most websites were static webpages, and the vast majority of users were consumers, not producers, of content.
Web2	Web2 has been around since the early 2000s when the emergence of large platforms like Facebook, Google, and Amazon, as well as services like Uber and Paypal, brought a centralized, commercial order to the internet by making it easier to connect, browse, interact, and make transactions online. These large companies capture much of the monetary value created on the internet.
Web3	Web3 is proclaimed by some as the future of the internet, where we partially return to the individualized utility of Web1, with the introduction of "state" (e.g., accounts) based on blockchain technology and digital tokens that can foster a decentralized internet. Rather than the large players of Web2 capturing the bulk of monetary value, Web3 replaces the centralized entities with decentralized networks that distribute the value to creators, users, and developers.

The definition of Web3 can differ by source; however, the following characteristics seem to be consistently present in some of the more popular definitions. While some of these features both deserve and have been the featured topic of entire books, what follows here is a short overview of each feature.

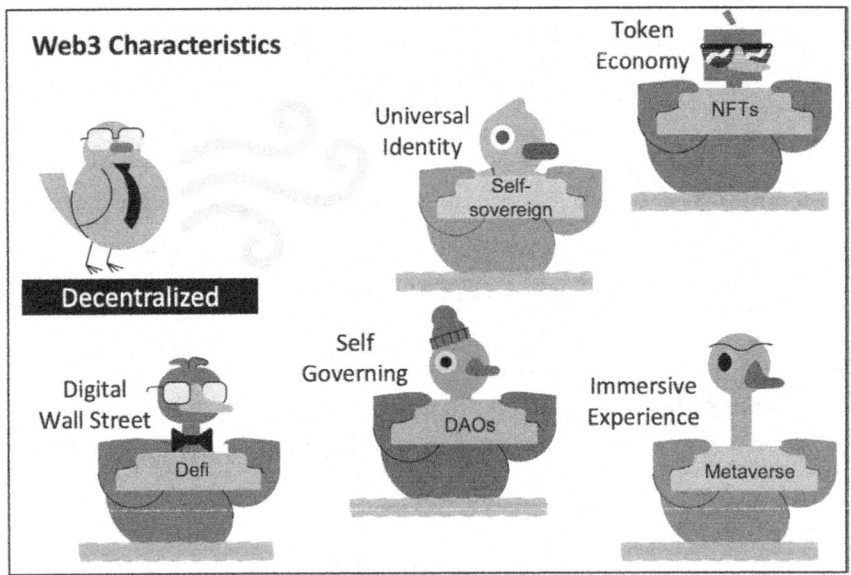

Chapter 9 – Figure 2 – Decentralization is the common theme for emerging Web3 features

Self-sovereign sign-on

Anonymous single-sign-on will allow one username and authentication method across all websites and accounts, rather than individual logins for each site. This login would not require you to relinquish control of sensitive personal data. This feature differs from current Facebook or Google single-sign-on, which grants access to your personal data until you revoke this access.

With Web3 wallets backed by blockchain networks this model is flipped around with the user always retaining control of their Personal Identity Information (PII) and login credentials. As crypto wallet services evolve, options will exist as to what type of blockchain network will back your wallet.

With a permissionless blockchain, all transactions on the blockchain are public, so technically, everyone can see the assets and data assigned to a specific wallet. This transparency is also why wallets are anonymous, identified only by an address, not a name unless the individual chooses to assign personal details to their wallets. Popular crypto-wallets include:

Coinbase	Coinbase wallet is often touted as the most popular wallet for buying and selling crypto-currency.
Metamask	Metamask crypto wallet & gateway supports a diverse set of blockchain networks and widely used as a wallet for holding NFTs and crypto-currency.
MyEtherWallet	An open source wallet popular with developers who are building digital assets on the Ethereum network
CryptX	An enterprise-grade operational wallet with strong security features. It supports over 100 coins and uses a simple interface to handle multiple coin balances.

A permissioned blockchain would allow a greater degree of privacy as to who can observe transactions between wallets, however your wallet service provider choices are much more limited because most of the popular wallets today are backed by permissionless networks. That said, remember when we covered digital identity in Chapter 8, we mentioned that your PII should not go on the blockchain? Today's wallets respect this notion, keeping PII locally (e.g., on your mobile phone) while the blockchain manages transaction history between wallets along with access permissions and rights information only.

Token-economy

Activities that contribute to Web3 are rewarded by a token (either NFT or fungible) to incentivize participation and distribute ownership. For example, when posting a new social message, an NFT representing that post could be "minted" (generated) and stored as an asset in a crypto-wallet. This token represents ownership over the message, which can then be traded with others via their wallets. If the post is popular, profits from advertising will go to the token owner rather than exclusively going to the platform it's hosted on.

NFT marketplaces are platforms where NFTs can be stored, displayed, traded and in some cases minted (created). These marketplaces are to NFTs what Amazon or eBay are to goods and are the seen as the epicenter for Web3's token-economy. Popular token marketplaces include:

Opensea	One of the most established universal NFT marketplaces. You can find Non-Fungible Tokens representing ownership of a wide variety of things, including artwork, sports collectibles, virtual worlds, trading cards, and domain names.
Rarible	An NFT marketplace owned by the community members holding RARI tokens. While it puts an emphasis on art, it also supports a wide range of other NFT items.
SuperRare	A marketplace focused on digital art that features a select handful of leading concept artists.
NBA Top Shot	An NFT marketplace for buying and selling digital collectible cards featuring videos of memorable NBA "moments."

Self-governing

Along with the distribution of ownership is the distribution of decision-making power. Without a central authority, blockchains rely on the entire network to verify an activity via consensus. However, specific systems, such as those used in Decentralized Autonomous Organizations (DAOs) can be established to democratize decision-making based on the quality or volume of a user's investment into a site or DApp.

For example, based on their share of ownership of a platform, users can vote on the rules that govern a site (e.g., maintenance schedules and fees). These rules are then executed by smart contracts. Notable DAOs include:

BeetsDAO	A community born within the larger EulerBeats NFT community that focuses on buying rights for music-based NFTs.
BitDAO	A large treasury fund that invests and governs across a broad range of DeFi projects with approximately $2.5 billion invested to date.
LexDAO	Legally-minded engineers, with crypto experience who to turn legal services into smart contract code. LexDAO headquarters is in the metaverse in the settlement of CryptoVoxels.

Immersive Experience

Metaverse is often linked to Web3 and similarly lacks a well-formed and agreed on definition. I can't help but think of the metaverse as a video game like Second Life or Fortnite, with an immersive experience similar to the one portrayed in the movie, *Ready Player Now*. While the metaverse is described to be far more expansive than a video game, the gaming world seems to have already adopted its most rudimentary form. However, given the decentralized theme of Web3, the metaverse claims not to be controlled by a single game company. Instead, it's an ecosystem the transcends those centralized applications, providing an open exchange of digital assets, including real estate and avatars.

Digital Wall Street

Decentralized Finance or DeFi is a broad term for Web3-based financial services anchored by public blockchains. With DeFi, you can do most of the things that banks support—borrow, lend, earn interest, buy insurance, trade derivatives & assets, and more. DeFi claims to be streamlined finance, eliminating paperwork, and greatly improving settlement speed. DeFi services are built around global blockchain networks including Ethereum and Bitcoin.

While DeFi services have yet materialize at scale, most advocates set a vision that expands on the notion of digital money and aims to create an entire digital alternative to Wall Street, but without all the associated costs including office towers, trading floors, banker salaries. The DeFi ideal has the potential to create more open, free, and fair financial markets that are accessible to anyone with an internet connection.

Connecting to the rest of the world

How do Web3-based smart contracts interact with the rest of the non-Web3 world? Oracles provide a way for the decentralized Web3 ecosystem to access existing data sources and legacy systems. Decentralized oracle networks (DONs) enable the creation of hybrid smart contracts, where on-chain code and off-chain infrastructure are combined to support advanced decentralized applications (DApps) that react to real world events and interoperate with traditional systems.[68]

When I think of oracles, I am reminded of how the Common Gateway Interface (or CGI-bin) revolutionized static Web1 servers to become dynamic Web2 servers. CGI-bin was later replaced by Java Servlets, which further enabled the static web to interact with the non-web world—including remote databases, mainframes, and other legacy software systems.

The blockchain oracle problem outlines a fundamental limitation of smart contracts—they cannot inherently interact with data and systems existing outside their native blockchain environment. Resources external to the blockchain are considered "off-chain," while data already stored on the blockchain is considered on-chain. Securely interoperating with off-chain systems from a blockchain requires an additional piece of infrastructure known as an "oracle" to bridge the two environments. Types of blockchain oracles might include:

Market Data Feeds	Pricing data that's been aggregated from hundreds of exchanges, weighted by volume, and cleaned of outliers and wash trading.
Environmental Data Feeds	Weather data and/or data coming from Internet of Things sensors that represent real-time data from external sources
News and Public Information	Real-time or historic information from the internet. Examples include sports scores, election reports and social media sentiment.
Any API	API calls to enterprise backend systems

- - -

While Web3 is currently a work-in-progress and isn't exactly defined yet, self-sovereign sign-on, tokens, DOAs, DeFi and the Metaverse each possess a strong aspect of decentralization. We have covered the topic of decentralization throughout this book. So, in a sense, while still under construction, you should feel confident that you're already familiar with many of the foundational elements of Web3: cryptography, decentralized cryptocurrency, smart contracts, tokens, self-sovereign identity and of course the underpinning of blockchain technology. For the remaining portion of this chapter, we will map out a potential road to Web3 by assembling the blockchain topics covered thus far in the book.

If a more decentralized internet is theme one of Web3, then the "stateful internet" is almost certainly theme two. The next section covers "state" and its relationship to Web3.

THE STATE OF WEB3

In information technology, a computer system is described as 'stateful' if it is designed to remember preceding events or user interactions; the remembered information is called 'the state of the system'.[69]

Today's internet is a state-less internet. Its participants can't hold their own state, nor transfer it from one to another, natively. Blockchains, starting with Bitcoin, gave us a way to have a stateful web of computers. Therefore, some Web3 advocates call Web3, the "stateful internet", or even "the internet of value" (I.e., in this case, state is where value is stored).

Web3 is an extension to Web2. Web browsers (e.g., Firefox, Chrome) and mobile devices (e.g., iOS, Android) are very much still relevant and the primary client interface to the Web3 world. However, new services that manage state are introduced, with blockchain networks at the backbone of the new state-full internet. As such, the architecture of Web3 applications introduce additional elements to the current Web2 framework, as well as new building blocks and tools for a developer to get familiar with. The following section dives deeper into how Web3 architecture extends Web2 as well as examines some of the top developer trends to pay attention to when developing against a Web3 design, where decentralized state management is key.

Web3 Architecture

I had the fortune in my career to be on the forefront of introducing Web2 to the industry. As one of the founding fathers of WebSphere Software, I did my share to bring Web2 to a list of the who's who of internet companies, including eBay, Paypal, ESPN/Disney (Go network) and many more.

WebSphere Application Server emerged as the poster child for the 3-tier client server architecture as shown in the figure below. This simplified view of today's Web2 architecture includes client software, usually running in a web browser or a mobile application, and a set of specialized servers providing content, logic, and data services. The flow, labelled with the number zero, shows a typical http request traversing from the web/mobile app, making three "hops" across each server type, and returning results to be displayed in the app.

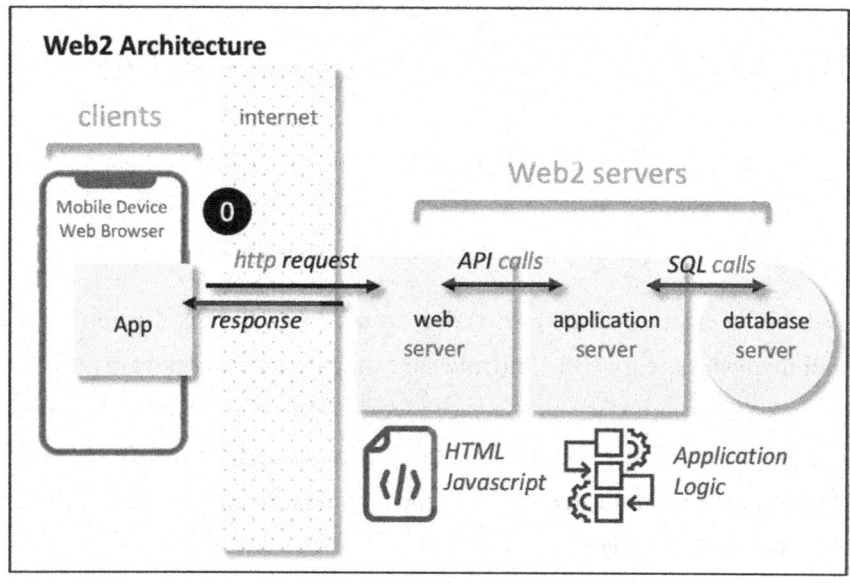

Chapter 9 – Figure 3 – Three tier architecture includes content, logic, and data services

In almost all cases, the three servers are owned, operated, and controlled by the same organization (i.e., company, government, school). Staying with the baseball theme set at the top of this chapter, let's consider this example. Imagine that a Big Baseball League runs their Web2 site using such an architecture. In this model, the Big League has sole control over who can access its servers' contents, logic, and data.

Furthermore, once a user creates an account and "clicks" to agree to their terms, the Big League can track user records, as well as change or stop their service at any time. Imagine the baseball league has a premiere fantasy baseball feature that allows you to create, trade and purchase rights of star players. As many fantasy leagues do, you can then play your fantasy team against others in the league and win prizes, etc. I have friends that put considerable time and money into such leagues. Now consider what might happen if the Big League changes its terms, or even goes away. As a user, you really have no power to preserve the value that you've created because your data is not your data.

Web 3 architecture builds off the Web2, by moving critical data outside the confines of any single organization and into a universal state management layer, residing on a blockchain. This architectural addition holds the promise to give the creator direct access to the enduring value that they've created. (Okay, perhaps my Fantasy Sport League is stretching it, but I think you get the point). However, as you see, the intermediaries (e.g., Big League) does not have sole control over the servers that hold your data. Hence a glimmer of hope exists to enable you to be in full control of your data.

The Web3 architecture can achieve these goals by:

Going Public - allowing applications to place some or all their content, logic, and data on to a public blockchain and decentralized storage networks.

Owner Consent - Now public and accessible by anyone, provided they have consent from the content, logic, and data creators/owners.

As implied before, Web3 does not replace Web2 as you can see in the figure below. Specifically, the figure represents a Web3 architecture and flow number zero shows how Web2 is still quite central to the overall design. However, there are two new infrastructure components that early Web3 applications are using to achieve its goals: **wallets** (flows 1 & 2), **blockchain gateways** (flows 3 & 4) to distributed blockchain and file storage networks. The following section provides a closer examination of these two components.

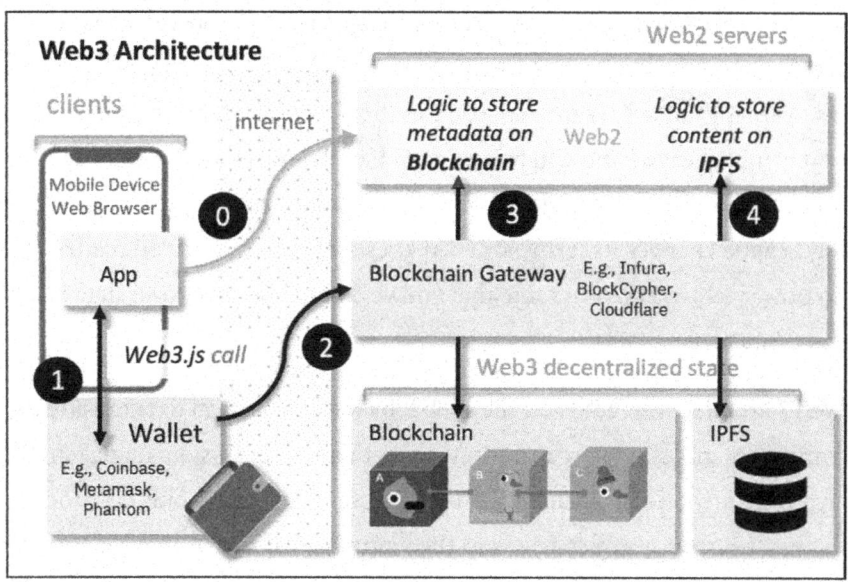

Chapter 9 – Figure 4 – Web3 architecture builds on Web2 adding a universal state layer

Wallets

Modern crypto wallets, such as Coinbase Wallet and Metamask interact with a web/mobile (client) app to allowing for a seamless user experience. Web3.js is a popular library used today by many developers to communicate between a client app and a user's crypto wallet. Web3.js library is designed to work with the Ethereum blockchain. A JavaScript code snippet to create a users' crypto wallet might look something like this.

```
 1  // start by loading the web3.js module;
 2  // allows us to use all its goodies
 3  const Web3 = require("web3");
 4
 5  // default provider
 6  let provider1 = "http://localhost:7545";
 7
 8  // use Infura as a blockchain gateway
 9  let provider2 = "https://mainnet.infura.io/v3/YOUR_PROJECT_ID";
10
11  // use Cloudflaire as a blockchain gateway
12  let provider3 = "https://cloudflare-eth.com";
13
14  // Use HTTP as the way to communicate with the gateway
15  let web3Provider = new Web3.providers.HttpProvider(provider1);
16
17  // Create the Web3 Object... and let the fun begin
18  let web3 = new Web3(web3Provider);
19
20  // create 1 account, with a wallet
21  let account = web3.eth.accounts.wallet.create(1);
22
23  // display address and privateKey to console
24  console.log(`address: ${account.address}`);
25  console.log(`privateKey: ${account.privateKey}`);
26
27  // Typical Console Output might look like this...
28  // address:
29  //0x854aea3473ba36bb56454A20741Daa8a04d29448
30  // privateKey:
31  //0x3b6f1424c5db5f84beb54fb8214aeb4408a2b71bb28f771ab30f7362497a2cc8
```

In the example above, line 3 is used to load the Web3.js library providing developers a rich set of functions. Lines 5-12 provide 3 options for accessing the Ethereum blockchain through a gateway (described more in the next section). Line 21 creates a new account with a wallet. And lines 24-25 print the address and private key of the account to the console as seen in lines 27-31.

Blockchain Gateway

In chapters 3 & 4, we discussed the multiple roles performed by nodes in a blockchain network. We defined that a miner is represented by a node in the Bitcoin (or Ethereum, or...) network that collects transactions and works to organize them into blocks. We also described how blockchain nodes are used to submit new transactions to the network. In fact, there now exist many different types of specialized nodes, such as full nodes, light nodes, archive nodes, mining nodes, and worker nodes. Depending on the requirements of a blockchain, each type of node plays different roles to secure and keep the network operational.

A blockchain gateway is designed to provide developers with easy access to decentralized blockchain and storage networks. Gateway services such as Infura remove the need to set up and oversee the management of your own node. Some even refer to these gateways, as a blockchain-as-a-service because it de-burdens the developer from obtaining, running, and operating their own node, allowing them to get right down to business and interact (e.g., issue transactions) to a blockchain.

Looking again at the previous architectural figure, we can see the blockchain gateway being used by the Wallet provider to submit a transaction to the blockchain (flow number 2) or query and add to state information on a blockchain (flow number 3). Gateways like Infura also offer developers APIs for storing and retrieving content from decentralized storage networks like IPFS (flow number 4).

Hyperledger FireFly is an open source project that provides a new type of decentralized orchestration layer between companies' existing systems and Web3 services. FireFly's SuperNode acts as a blockchain gateway with a simplified developer interface to write DApps once, then run and deploy them on a choice of blockchain networks. Kaleido's SuperNodes-as-a-service is an easy way for developers to experiment with Web3 architecture and is part of the Think Blockchain Labs Assignment for this Chapter.

- - -

Again, Web3 does not replace Web2 it extends Web2 with decentralized state. This enables the "off-boarding" of key personal and application data from centralized organizations to decentralized wallets, blockchains and storage networks where the user retains control and ownership of the value that they've created.

Now that we've introduced the system architecture of a Web3-based internet, we are now ready to look closer at Web3 from a developer's lens.

Web3 Developer Experience

Tools and frameworks are a developer's best friend. I am envious of the wealth of tools and purposed frameworks that are available to Web3 developers versus the tools sets that I used when doing my share to establish Web2. In fact, the Web3 developer tool market is booming making it difficult to pick a starting point. However, I found that the Ethereum development ecosystem seems to be the most active. The following table outlines some of the tools available to Web3 developers that are building against the architecture described in the previous section.

Although I won't go through the tools available in detail, it's helpful to know which ones are at developers' disposal. The list below is by no means comprehensive but includes tools new Ethereum developers often start with. Remember, building a Web3 app means you are also building a Web2 app, so don't read the columns below as Web2 "or" Web3, read them with an "and".

Tools Type	Web2	Web3
Proxy	Apache Web Server	Geth
Back-end services	AWS, Azure, IBM Cloud	Infura, Kaleido
Integrated development Environment (IDE)	VSCode	Remix

Application server	nodeJS, WebSphere Liberty, JBoss	Truffle
Front-end services	React	Web3.js
LibraryRepo	Npm, pip	OpenZeppelin
Testing Platform	Test IQ, Lambda Test	Ganache
Security Verification	AppScan	Quantstamp

Decentralization design choices

To decentralize or not to decentralize, that is the question. I have noticed that some Web3 developers are overzealous here and are gunning to put it all on the blockchain. Patient grasshopper. We might get there someday. There is much to consider here especially given the slow and expensive nature of present day blockchains. The popular CryptoKitties were perhaps the first DApps that tried to keep certain portions centralized. For example, their breeding logic is not public. Although they have received some criticism for this, it hasn't stopped users from spending a significant amount of money to purchase cats bred by this logic.

A first principles way of approaching this choice would be to adopt a "minimally viable public state" approach.[70]

For example, if you are building a fantasy sports league where users can own assets, then ownership should be on the blockchain. Also, the accounting for the reward tokens for winning should be on the blockchain. At the end of the day, users will find your application valuable if they can claim true ownership over the key activities your application enables.

- - -

Perhaps it's worth repeating just one more time. Web3 is an emerging ideal that will crystalize over the decades to come. There are many aspects that are still quite aspirational. And many that have shown early signs of life, require quite a bit of attention. The next section covers some of the top areas that require attention and/or where major questions still exist. This might be the most important section of this chapter because it's a call-to-action of sorts about the work left on paving a road to Web3.

STILL A WAY TO GO...

To quote Frank Costanza, from the *Seinfeld* TV show regarding the tradition of Festivus, "It begins with the airing of grievances. I got a lot of problems with you [Web3] people and now you're gonna hear about it."[71]

In this section, several yet to be solved problem areas are enumerated and discussed. I try to take a balanced view and not only air the grievances, but also try to propose areas to investigate that might remedy the problems.

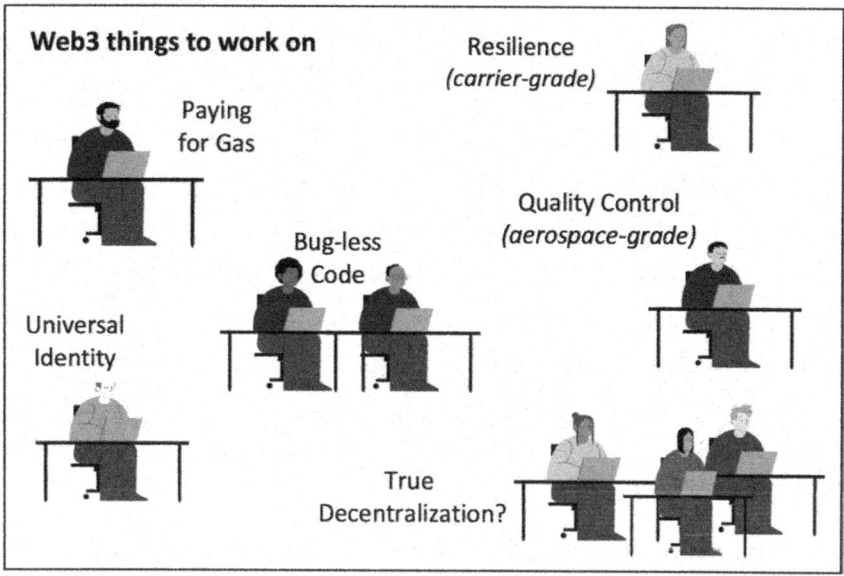

Chapter 9 – Figure 5 – Web3 problems yet to be solved

Paying for Gas

Powering smart contracts on public networks is becoming increasingly prohibitive. For example, every DApp built on Ethereum today requires virtual gas to power transactions, which is paid for in ether (ETH). Low-cost NFT adoption can't survive on the Ethereum blockchain with triple figure gas fees. This won't be feasible over the long- term, if millions of non-crypto-native people are to use Web3 applications.

"High gas prices" can be tempered by more competition including from private permissioned blockchain networks. If ether is too expensive, why not run your contracts on Solona, Hedera or Cardano? However, it's not that easy. Most DApps today are written to work against a specific blockchain network and do not support a "write-once, run-anywhere" architecture. And once your data is on the chain, how easy is it to move to another chain?

For now, my suggestion is to be highly selective about what gets put on the blockchain. Also, there are developer frameworks similar to Web3.js that support multiple blockchain networks. For example, Tatum, provides a JavaScript library supporting over 40+ blockchain networks. While a framework like Tatum helps, the problem of migrating data from a high-cost to a reasonable cost chain has not yet been perfected. This is work still left to be done.

Universal Identity

As we've discussed several times already in this book, self-sovereign identity is key to improving our state of cybersecurity as well as how we support Web3's "your data is your data" mantra. Since there aren't many functional decentralized identity solutions today, some DApps are still asking users to create an account, enabling identity information to be associated with their activity on the app. This is not too different than the Web2 way of doing things. Once we have functional decentralized identity solutions, how should DApps treat and present this?

Although there's no clear answer, the Ethereum ecosystem has proposed ERC-725. 725 is an open, portable standard for identities that enables decentralized reputation, governance, and more. Users will be able to take their identity across different DApps and platforms that support this standard.

Hyperledger Aries also holds promise to contribute to solving this issue. Aries is a lower level set of developer tools for storing and exchanging data related to blockchain-based identity using W3C standards. The existence of tools like Aries lowers the barrier for developers who are determined to evolve identity beyond centralized Web2 architectures. This is work still left to be done.

Bugless code

Hah! There is no such thing as bugless code. Software is written by humans. And humans do make mistakes. However, smart contracts that power Web3 DApps, DeFi services and NFTs, need to be flawless, especially on public networks.

As discussed in Chapter 5, the 2016 DAO incident was an exploitation of a bug in the DOA smart contract. In fact, as I was writing this book, I read a news article claiming the National Basketball League, had just launched its first set of NFTs related to their 2022 playoffs. Within minutes of launch, the Ethereum smart contracts underpinning the NFTs were hacked and $68,000 was drained. The NBA received quite a bit of grief that despite annual revenues of $10 billion, they apparently could not afford sufficient code reviews or bug bounties to keep the NFTs safe. The more likely truth is, the NBA probably didn't know any better.

Better the devil you know than the devil you don't know – might be a good way to describe the vulnerability of Web2 (known) versus Web3 (less known). Web3 companies must employ Web2 code security and review practices. This includes peer reviews and code bounties.

Three considerations come to mind as we strive towards a bugless Web3 world.

Perfect code - The "perfect code" problem seems to be indigenous to permissionless networks. In a permissioned network, there is a much greater chance that a would-be hacker would be known to the network. One might reason that permissionless networks should not yet be considered for digital assets of extreme monetary value. One argument against that is Bitcoin, which has been hacked far less then Ethereum. However, Bitcoin is a fixed function network (cryptocurrency) making its attack surface one-dimensional. Ethereum has a multi-dimensional attach surface thanks to its powerful support for smart contracts and the token economy that follows. Hence, is there a rule of thumb that says if the digital asset in question is of extreme value, either mint it on a permissioned network, or "one-dimensional" networks akin to Bitcoin?

Just keeps running - The next consideration has to do with governance of the network and code. Outside of bugs, code is expected to evolve with new innovative features. This includes both the code represented by smart contracts as well as the code powering the network protocol itself. For example, it is impressive to see in the telecommunications market how wireless protocols from 3G, LTE and now 5G have been introduced to the market (and in 3G's case deprecated) without interruption and without the network skipping a beat. Network protocols as with internet and world-wide web protocols are all forward and backward compatible. This is where the term "carrier grade" comes from, referring to hardware or software system that is extremely reliable, well tested and proven in its capabilities. Carrier grade systems are tested and engineered to meet or exceed "five nines" high availability standards, and provide very fast fault recovery through redundancy (normally less than 50 milliseconds). Code version-upgrades/downgrades and patches are a fact of life in any software system. A resilient design with tight governance of code will be expected by Web3 users to be no less than carrier grade.

Quality with a capital Q – Beyond carrier grade there is aerospace grade. Web3 users will demand we treat the quality of smart contracts with the same quality control that the airline industry treats creating airplanes. Airline manufacturers (or spacecraft, nuclear power plants, etc.) have extreme quality control over their engineering processes. They invest in flight simulators which have evolved to be accurate digital twins of the real crafts. And when the plane is built, test pilots run it through a rigorous set of flight plans. Data is gathered, analyzed and defects are brought back to the drawing board. But the industry does not limit its quality control to any single vendor. The Federal Aviation Administration (FAA) oversees operations across all vendors with a goal to provide the safest aerospace system in the world. Does Web3 need an equivalent to the FAA? Think about it. If there is a bug in an airplane (i.e., a crash), not only does the FAA work with the vendors to debug the issue. They also write guidelines and policy that all vendors must follow when incidents are found. Too much oversite? I don't know. I'm not sure I would step foot in a plane if it weren't for the FAA. So, would you write an NFT or smart contract that puts your family's life savings at risk without similar oversite? Hmm. Makes you think. Certainly, something that needs to be considered as we pave a road towards Web3.

Part 1 – Is there such a thing as true decentralization?

I have yet to meet a truly decentralized system. Bitcoin comes close. Ethereum, less so. And while we can debate how decentralized these systems are, the utilities built around their ecosystems are rarely decentralized. And if you are as strong as your weakest link, then I would venture to say, these networks that are often touted as decentralized, are not as decentralized as you might think. Let's look at a few examples.

My dad uses Coinbase.com as his crypto exchange. It's where he keeps his wallet and his account. The user experience, especially of their mobile app is exquisite which is what keeps people coming back to make frequent use of it. However, Coinbase is a centralized company, which is a gateway to multiple decentralized crypto networks. However, if Coinbase goes down, my dad would not know how to get to his Bitcoin account, even though Bitcoin does not go down. However, the way he describes it is, "Bitcoin is down."

Similarly, in a previous section we discussed the ease-of-use that blockchain gateways like Infura bring to developers. The Ethereum developer ecosystem depends on it, and if you've ever used its APIs, you would know why. However, several times Infura has gone down, like it did in November of 2020. Given Infura acts as key infrastructure for the Ethereum network, its crash caused problems for users on Binance, Metamask, OpenSea and other popular apps. Ethereum was not down, but users' access to it was. Hence, as my dad would say, "Ethereum is down."

Part 2 – Is there such a thing as true decentralization?

Part 1 points out the irony of decentralization in the light of the various centralized systems that people depend for access. Part 2 poses the question – how much decentralization do you really need?

Decentralization is critical to ensure trust. As the more a system is decentralized away from a single entries control, the less likely that entity can tamper with the integrity of that system. So, how much is enough? Let's look at an example.

An article published by Duke University explains.

A jury of 12 resonates through the centuries. Twelve-person juries were a fixture from at least the 14th century. Over 600 years of history is a powerful endorsement. So too are the many social-science studies consistently showing that a 12-person jury makes for a better deliberative process, with more predictable (and fewer outlier) results, by a more diverse group that is a more representative cross-section of the community.[72]

If twelve is a tried and true number for gaining trust in judicial systems, why isn't the number good enough to verify transactions on a decentralized blockchain?

Based on my work on permissioned blockchains with IBM, I've asked and been asked this question. Most of the examples shared in Chapter 6, including Food Trust, Trade Lens, and Verfied.me are all decentralized but have finite number of nodes in their network—and in most cases the number is less than 20. This has gotten me to think about how public blockchain networks might be created with a fixed number of diverse nodes. Let's face it, most of us trust our bank, telco, city, university, government, etc. This type of trust is almost always grounded in reputation. These institutions have demonstrated consistent behavior, in some cases over decades. Let's call this type of trust reputational trust. And if they let us down, that's okay, we have freedom of action and can take our business elsewhere. Blockchains' cryptography, append-only transactions and distributed nodes, form the foundation of algorithmic trust.

So, rather than have expansive and completely trustless public networks, perhaps we can present a similar outcome with a combination of reputational and algorithmic trust. This would be an alternative approach to solely building our networks on algorithmic trust which is how most public blockchain networks operate today.

It might work something like this. We have an election system to pick twelve (think jury) willing participants to manage nodes in a local, regional, continental, and/or global blockchain. There are the blockchain governors. We assume these twelve governors were elected because of their strong positive reputation. We might even consider the twelve to be a DAO of sorts, who set the rules for operating the network including how new members to this DAO get elected. An incentive system would compensate the governors for operating nodes. No mining or proof of anything needed to operate. Okay, I'll admit the details are thin here, but hope you see where this thought can go. Also, as I mentioned, I have seen dozens of networks form from such governance principles under the Hyperledger ecosystem.

In conclusion, it's not clear to me that such an approach would be any less decentralized (for reasons cited above) then what we have today. I will keep thinking about this one, and hope you do to.

THE INTERNET OF THE FUTURE OR A BUZZWORD?

The Web3 vision is of a future, decentralized phase of the internet, where users become owners. While this might seem revolutionary to some, the example of the Players Baseball League demonstrates that the issue of centralized control has been an issue that society has been successfully addressing since the 1800s.

It would also be fair to say many, including the author of this book (yours truly) are skeptical about how decentralized we can or want to become as a society. That said, this chapter covered a balance perspective of how Web2 applications could be enhanced by off-loading personal data to a universal state layer involving decentralized wallets, blockchain and storage networks.

While Web3 builds upon Web2, and is still in its infancy, its early examples paint a picture of great potential benefits over the current internet. However, its list of open issues can make it seem decades away. Although Web3 has yet to crystalize enough to capture major market share, interest from investors and venture capitalists is likely to grow as people become more educated on this new phase of the web.

QUIZ TIME

1 – Which best represent the characteristics of Web3?

    a. Decentralized internet

    b. Stateful internet

    c. Internet of Value

    d. All of the above

2 – Which capabilities are grouped under the Web3 movement?

    a. Unified Identity managed through crypto wallets

    b. DeFi services built atop cryptocurrency networks

    c. NFTs reward Web3 contributing activities often governed by DAOs

    d. All of the above

3 – Web3 architecture introduces a universal state layer through these services?

    a. Wallets, Blockchain Gateways, Decentralized Blockchain, and Storage Networks

    b. Web2 content, logic and data servers

    c.   Client apps that run in browsers or mobile devices

    d.   None of the Above

4 – Gas prices, Adoption of universal identity, bugless smart contracts, and pursuit of true decentralization are inhibitors to Web3's full adoption?

    a.   True

    b.   False

5 – Think Blockchain Labs– Chapter 9 Assignment

Go to the Think Blockchain Labs Github repository.

See the README.md file and search for the Chapter 9 assignment details, but the gist of the assignment is to:

    a.   Run the Web3.js example as illustrated and discussed in this chapter

    b.   Build a Web3 blockchain DApp using Hyperledger FireFly using Kaleido's cloud service

UP NEXT

The path forward described is an attempt to provide a balanced view of how the internet will evolve. Decentralization, shift of value to creators, a stateful internet augmented with blockchain are goals worth striving for. However, the ultimate goal, as shown through examples in this book, is to change life for the better. Eliminate identity theft, foodborne illnesses, counterfeiting and reducing cyberattacks are all examples of a better life and my hope for what Web3 will deliver. As illustrated in the following right-most-photo from my camera roll, I hope to continue to do my share to educate how blockchain can "change everyday life for good."

The photo on the right is of my friend and colleague Nitin Gaur, who is a Web3 subject matter expert on all things DeFi. Listening to Nitin on podcasts is one way I keep informed and energized on Web3 topics.

Chapter 9 – Figure 6 – Web3 aiming at the idea of changing everyday life for the better

- - -

Winding down, this next chapter provides a summary of the books key takeaways, ties up some loose ends, and shares a few options on how to stay in touch.

Chapter 10 –
KEY TAKEAWAYS

Wow... you're still here? I applaud your perseverance. Well, since you're still reading, let's share a few closing thoughts and summary.

COVERED IN THE CHAPTER

- Key points to remember

- Why the ducks?

- See you in the Lab

CIAO

The word "ciao" is mostly used as "goodbye" or "bye" in English, but in modern Italian and in other languages, it may also mean "hello." Ciao captures the spirit of this short chapter. It is a bit of a "bye for now" while also providing a "hello" and a means to stay in touch. Playing on that sentiment, this chapter provides a summary of the books key take ways, ties up some loose ends, and shares a few options on how to stay in touch.

KEY POINTS TO REMEMBER

You've seen a few key thoughts repeated in this book. This section summarizes the most essential points to remember from *Think Blockchain*.

If databases were birds, blockchain would be a duck

This statement is a fun, and maybe even slightly bizarre attempt to make a simple point. Blockchain technology has many similarities to database technologies. And if you understand the main attributes of a database, you can build on that as you venture into learning about how blockchains work. Hence, you wouldn't to be completely right to say a blockchain is a database, but you wouldn't be completely wrong ether. (A little joke; ether, get it?) So, when introducing the topic of blockchain technology, I find it helpful to start by finding a common point of reference, in this case it's by likening blockchain to databases and building from there.

The first key point worth remembering is that blockchain's truly brilliant architecture is built on decades-old fundamental research in cryptography, distributed transaction processing, game theory, and yes, database technology too.

Bitcoin is just one of one thousand applications of blockchain

As you've heard me say, if there are one thousand applications of blockchain technology, Bitcoin is only one. In fact, the focus on this book has been to look broadly at how the evolution of blockchain technology enables the construction of a new and broad class of applications.

Now in all fairness, Bitcoin gets the credit for coining (no pun intended) the term 'blockchain' and demonstrating its abilities on a global scale. However, blockchain is a technology with many varieties and implementations. In this book, we covered the history of blockchain by examining three pioneers of blockchain technology. Each pioneer paved the way by introducing a game-changing concept that stretches the state-of-the-art of distributed ledger technology. This included:

- **Bitcoin** as an introduction to **cryptocurrency** and the breakthrough technique of **mining**.

- **Ethereum** evolving blockchain to being a world computer for creating **digital assets** (tokens) with the breakthrough of **smart contracts.**

- **Hyperledger Fabric** further evolved blockchain for the **enterprise**, with a Lego-like **modular** architecture that enables pluggable permissioned identity, privacy and new scalable consensus models.

Blockchain technology comes in many shapes and sizes. Modern blockchain is flexible and ultra-efficient. Oh… and it doesn't always consume energy like Bitcoin does. The expanded versatility of blockchain technology also expands the scope of applications that can be built. With that, there are few barriers to prevent blockchain to be applied liberally across all industry use cases.

Therefore, the next key takeaway is that applying blockchain technology broadly will transform industries—because its application can go far beyond cryptocurrency and Bitcoin. Here is a partial list of possible applications, which was introduced back in Chapter 1, that I hope motivates you to think deep and wide about where to apply blockchain.

- Registries for land, vehicles, guns, etc.

- Medical data and clinical trial marketplace

- NFT marketplaces

- Music rights and royalties tracking

- Cross-border payments

- Supply chain, logistics monitoring and dispute management

- Voting

– just to name a few.

Decentralized blockchain establishes unprecedented trust

Decentralization refers to the transfer of control and decision-making from a centralized entity to a distributed network. Decentralization of data via a shared ledger is one of the keys to its trustworthiness. For example, decentralization makes the Bitcoin network secure and difficult for malicious people and hackers to penetrate or interfere with because they can't control the entire system.

Blockchain's power to transform is that it enables co-development of a shared copy of the truth. And with this, what a group can achieve together far exceeds what any individual member can achieve by themselves. Therefore, I like to say **blockchain is a team sport.**

The decentralization topic is a repeated theme throughout many chapters in this book. Ethereum's notion of a world-computer forms the foundation for decentralized apps (DApps). The token economy, NFTs, self-sovereign identity, Metaverse, decentralized finance (DeFi) are all excellent examples of blockchain-based applications that leverage decentralization.

The Cybersecurity Chapter emphasized that blockchains possess ingredients that make it more naturally resistant to cyberattacks. While the attackers are decentralized, the targets of cyberattacks are largely more centralized. The inherently decentralized nature of blockchain technology can level the playing field against hackers.

Finally, the Web3 chapter paints a picture of a future state of the Internet where many of its most basic services become decentralized. However, the Web3 chapter also lists some issues to overcome with the decentralization concept. The chapter strongly suggests we look at a balanced view of decentralization.

The notion of designing for **Minimally Viable Decentralization (MVD)** is the key skill to learn for Web3 application development. Building on MVD, we also posed the question of "how much decentralizing is good enough to establish trust?" Many of the permissioned blockchain networks created with Hyperledger technology follow the rule of twelve, which seems to work on building trust within our legal systems (i.e., a jury of twelve).

Blockchain-based trust breeds widescale adoption
Let's face it, most of us trust our bank, telco, city, university, government, etc. This type of trust is almost always grounded in reputation. Blockchains' cryptography, append-only transactions, and distributed nodes, form the foundation of algorithmic trust.

Some people call blockchains **trustless**. However, the term 'trustless' is a bit confusing to me because blockchains don't actually eliminate trust. What they do is minimize the amount of trust required from any single actor in the system by distributing trust.

So, rather than have expansive and completely trustless public networks, perhaps we can present a similar outcome with a combination of reputational and algorithmic trust. This would be an alternative approach to solely building our networks on algorithmic trust which is how most public blockchain networks operate today.

Another key takeway is that blockchain's combination of reputation trust and algorithmic trust calms fear, trepidation, and encourages widescale usage from a large diverse population of users. When this trust model is applied correctly a safe, secure, and vibrant ecosystem will quickly emerge.

Blockchain at the nexus of emerging technology

In Chapter 7 we covered the notion of blockchain at the nexus of technology, which includes artificial intelligence, the Internet of Things, and quantum computing. In each of these cases, blockchain technology can introduce the missing element of trust into the picture. This includes using blockchain as an audit trail for training data of machine learning models and IoT firmware versions allowing users to quickly identify whether associated data is authentic and has not been tampered with.

The key take way here is trust is also essential for the wide-scale adoption of emerging technologies. With software and open source "eating the world" trusting technology is not an option, it is essential. From trustworthy AI to protecting our software supply chain using SBOMs, blockchain technology acts like a "food label" with an accurate, tamper resistant log of the key ingredients, history of modifications and providence of ownership. This level of rigor of making our technology known and transparent, will accelerate adoption and hyper-growth for our digital economy.

Web3 is the future internet fueled by blockchain

This takeaway is regarding Web3, which many say is the future of the internet. Skeptics say, it's just a buzzword. However, there is no doubt in my mind that the wonderful thing that we call internet will evolve. The next wave must tackle the toughest challenges covered in this book, however, doing so will provide a means to flow more control and rewards to those creating the value on the web.

Web3 architecture builds on Web2 and introduces a universal state layer that allows creators of data to own, control, and monetize their data. As such, Web3 ushers in a labor shift, like ones we've seen in the media and entertainment industry, from exclusive control by owner, to shared control with the data creators. This behavior can be seen in Web3 applications that we studied in Chapter 9 including NFTs, DeFi, Metaverse and Universal Identity, to name a few.

However, Web3s list of open issues can make it seem decades away. This includes issues with ensuring code quality, matching up to carrier/aerospace-grade networks, cost of gas, and striking a balance on decentralization.

Although Web3 has yet to crystalize enough to capture major market share, interest from investors and venture capitalists is likely to grow as people become more educated on this new phase of the web.

Blockchain is changing everyday life for good

How many of you believe, like me, that blockchain IS changing everyday life? Alright, perhaps you think it's too soon to make such a bold statement. Well, as this decade rolls on we are seeing some compelling evidence that this is true. In this book, I shared several examples of blockchain applied for good. Examples covered includes eliminating food-borne illness with the Food Trust network, giving user control of their digital identity to end identity theft with the Verified.me network, and reducing counterfeiting with IBM Crypto Verifier.

Like many movements in the industry, there are always extreme points of view. Extreme views on blockchain tout it as a means to eliminate intermediaries like big businesses and governments that unfairly dictate control over the world's population of digital users. In this book, I've tried to take a balanced view. Specifically, I feel strongly that there is a necessary role for all governments and businesses (big and small) to form a safe and prosperous future state for our Web3-based internet.

The final key takeaway is that blockchain is a technology that is poised to change everyday life for the better. It is here for good. The double meaning is that it is here to stay and to bring a new level of trust, built on reputation and algorithms. A balanced view will almost certainly prevail.

Hybrid blockchains are the responsible choice of network designers because they often possess a minimally viable decentralized design and are deployed to an interoperable web of public-permissioned networks. And yes, changing life for the better also means it's time the creators of data—whether that be an NFT, your healthcare record or your digital identity—be granted control and profit from that data.

Therefore, the responsible development of blockchain is a team sport involving users, governments and businesses collaborating to both protect and liberate the world's data. With this, blockchain as a technology and a mindset breeds trustworthy computing and becomes the fuel that propels Web3 and the future of our internet—for good.

WHY THE DUCKS?

Honestly, I really don't know. And when you come right down to it, I see myself as more of a dog person. But certainly, birds and ducks are okay too. I guess it was fate that made blockchains into ducks, versus eagles or Chihuahuas. And Shaun Lynch and I certainly had fun littering this book with playful illustrations of ducks. I hope you also had some fun with the quirky yet informative illustrations. I know I can't help but smile every time I see a duck and will always Think of Blockchain when I do.

SEE YOU IN THE LAB OR LINKEDIN OR IN CLASS

I'd love to stay in touch with you, get your feedback, and expand on the examples that we covered in this book. Your feedback and extra-curricular coding may very well make it into the next edition of this book.

The best way to connect and/or share feedback is to send me a message either on LinkedIn or to my email forwarding address.

- LinkedIn: https://www.linkedin.com/in/jerry-cuomo/

- email: jerry@thinkblockchain.org

To engage with code, go to our Think Blockchain Labs repository on GitHub.

- Github: https://github.com/JerryCuomo/ThinkBlockchain

- The README.md file contains the chapter by chapter instructions on how to run the samples and provides "challenges" for extending the samples.

Last, but not least, I hope to use this book as the teaching foundation for a university course on Blockchain. NC State University in North Carolina has been kind enough to offer me a chance. Once started, I hope to find a productive way to publish these courses on-line. Message or email me if you're interested in taking the "Think Blockchain" class. But given you already have the book; you have one foot firmly planted in the right direction!

UP NEXT

Well, I hope you've enjoyed the content in the book at least half as much as I've enjoyed writing it. Up next are a set of supplemental chapters that I thought would be an informative and fun way to wrap-up the *Think Blockchain* book. The first is a collection of my blockchain-related testimonies to the US government. As I reread these testimonies, I realized that the thoughts contained within were a useful takeaway and could act as an informative summary for this book.

Addendum 1 –
My Testimonies
To Congress

This supplemental chapter contains three transcripts from my blockchain-related testimonies to the US Government. Two of the testimonies were to congressional committees hosted by the House Energy and Commerce Subcommittee. The other was a President Obama initiated Commission on Enhancing National Cybersecurity.

People sometimes ask me, why were you asked to do this? Did you do something wrong? No, I did nothing wrong. In fact, it was quite cool. You see, IBM has a good relationship with the US Government and tries to do its share to educate on matters that involve technology. Given my work with open source blockchain and our world-wide customer experience with blockchain projects, was likely the reason why I "got the call." During my first testimony, a congressman pulled several of us aside and asked if we knew why we were here? This was his successful attempt to calm our nerves. He said you are experts, that's why you are here. He added, members of Congress would like to learn from you, and we will likely have questions. Lastly he said, think of this like serving jury duty. It's a form of community service. I must admit, his words did calm me down for about five minutes. But the butterflies in my stomach quickly returned when I stepped into the large chamber where the congress-folk were. All-in-all, a wonderful experience that I won't soon forget. Enjoy reading the transcripts.

TESTIMONY 1 – MARCH 16, 2016

House Energy and Commerce
Subcommittee on Commerce, Manufacturing & Trade
How to Capitalize on Blockchain
Gennaro (Jerry) Cuomo, IBM Fellow

Good morning, Chairman Upton, Ranking Member Pallone, Chairman Burgess, Ranking Member Schakowsky, and members of the subcommittee. My name is Jerry Cuomo and I am IBM's Vice President for Blockchain Technologies. Thank you very much for the opportunity to testify this morning.

Technology and business leaders at IBM believe that blockchain is a revolutionary technology. It's a foundation for building a new generation of applications that establish trust and transparency while streamlining a wide variety of transactional processes. You are wise to include blockchain in your study of "disruptive" technologies because blockchain has the potential to vastly reduce the cost and complexity of getting things done—across industries, government agencies and social institutions.

I also want to tell you what blockchain is not—it's not Bitcoin, the cryptocurrency. While blockchain is the core technology that enables Bitcoin to operate, it can be used for entirely different purposes. Whereas Bitcoin is an anonymous network, blockchain can be used to set up trusted networks to handle interactions between known parties.

In this paper I'll explain what blockchain is, how it works, how it can best be built and used—for the benefit of business, the economy, and society.

Key points:

Blockchain creates trustworthy and efficient interactions. It's a distributed ledger shared via a peer-to-peer network that maintains an ever-expanding list of data records. Each participant has an exact copy of the ledger's data, and additions to the chain are propagated throughout the network. Therefore, all participants in an interaction have an up-to-date ledger that reflects the most recent transactions or changes. (The "block" is the record and the "chain" is the collection of blocks that populate the ledger.) In this way, blockchain reduces the need for establishing trust using traditional methods.

Blockchain technologies must be enhanced to meet the needs of businesses. The core technology must be adapted to further address security and privacy concerns—creating an enterprise-ready blockchain. In addition, computer systems and networks must be architected so they can scale up to handle an immense volume of transactions and industries and governments begin using the technology to handle their core organizational processes—and complete their tasks in seconds rather than minutes.

Blockchains must be open and interoperable. For blockchain to fulfill its full potential, it must be based on non-proprietary technology standards to assure the compatibility and interoperability of systems. Furthermore, the various blockchain versions should be built using open source software, with a combination of liberal licensing terms and strict governance, rather than proprietary software – which could be used to suppress competition. Only with openness will blockchain be widely adopted and will innovation flourish.

Blockchain will greatly benefit from government participation. It's critical from a national competitiveness point of view for US companies and government agencies to lead the world in understanding the potential of blockchain and putting it to use. Because of the transparency made possible by blockchain, government agencies will be able to understand better what's going on within financial and commercial systems—and spot potential problems before they become critical. Blockchain will also enable more efficient interactions between government and businesses—regarding everything from taxes to land use.

Part 1: How Blockchain Can Be Used

Over the past two decades, the Internet, cloud computing, and related technologies have revolutionized many aspects of business and society. These advances have made individuals and organizations more productive, and they have enriched many people's lives.

Yet the basic mechanics of how people and organizations forge agreements with one another and execute them have not been updated for the 21st century. In fact, with each passing generation we've added more middlemen, more processes, more bureaucratic checks and balances, and more layers of complexity to our formal interactions—especially financial transactions. We're pushing old procedures through new pipes.

This apparatus—the red tape of modern society—extracts a "tax" of many billions of dollars per year on the global economy and businesses.

What can be done? Businesses, governments, and other institutions can use blockchains to build and govern business networks.

Blockchain-based systems could help radically improve whole industries, beginning with banking and insurance. But its impact could be much broader. It could make a difference whenever valuable assets are transferred from one party to another and whenever you need to know for certain that a piece of digital information— anything from electronic artwork to the terms of a business agreement—is unique and unchangeable by any party without the agreement of all parties.

I want to add a note of caution, however. Blockchain isn't the answer to every process or transaction-related problem. There will be situations, where it will improve efficiencies and provide other benefits, but there will be others where it's not a good fit. Furthermore, don't underestimate the technical and organizational challenges of building and adopting blockchain-based systems.

Here's where blockchain fits well, managing a business agreement between two or more companies. They can record the terms of that agreement on a blockchain, knowing it will execute and be enforced autonomously (e.g., "If you pay me in under 15 days, then I will give you a discount."). Nobody is in private control of the ledger and nobody can secretly change the terms of the agreement. It's like every guest at a B&B writing in the guest book with an indelible Sharpie. So, with blockchain, facts and agreements are recorded certifiably and indelibly, increasing trust, reducing risk, and thus reducing friction in business.

There's a broad range of potential business solutions. On one hand, enterprises will be able to reimagine well known business processes and areas like supply chain, securities trading and logistics. At the same time, blockchain is poised to enable enterprises and whole industries to invent new digital business processes that include connected devices (Internet of Things) like cars, smartphones, appliances, solar energy panels, and drones. This capability could be critical, for instance, in enabling the insurance industry to design liability insurance policies to cover autonomous vehicles.

IBM is already begun deploying a blockchain-based system internally—for managing our commercial financing business.

The financial services industry is in the forefront of blockchain adoption. Almost every transaction in financial services involves multiple parties and many steps, largely because of the checks and balances that are required to assure that what has been promised has been done. Consider how the technology might be used in a critical financial services process, the settlement in securities trading. People in the industry are talking about a concept they call T+0, which means same day settlement. The hope is that they'll be able to use blockchain to strip out the inefficiencies and handoffs that are required to settle a trade so that settlement occurs on the same day as opposed to 2 or 3 days later as it is today, depending on the market.

Now, imagine supply chains where blockchain is put to work. An aircraft manufacturer, for example, might create a blockchain-based system for holistically managing all of its relationships with suppliers of parts and components. All of the suppliers will share the exact same information about a new aircraft model—every step in the process of planning, designing, assembling, delivering, and maintaining it. At the same time, the manufacturer will use other blockchain-based systems for managing the financial relationships and transactions connected to each step. Thanks to blockchain, trust and accountability are built into supply chains. So are compliance with government regulations and internal rules and processes.

Blockchain fundamentally changes the game across three dimensions: time, cost, and risk. It reduces the time required to settle a multi-party contract from days to seconds, potentially. It reduces costs by stripping out intermediary organizations and processes. And, by enabling permissioned networks to share a transparent and non-changeable ledger, you reduce the risk of tampering, fraud and collusion.

Part 2: How Blockchain Works
Blockchain is both a software technology and a mechanism for groups working together.

At the heart of the blockchain network is a shared ledger, which describes assets, identifies their owners, lays out the steps in a process and records when each step is completed. Only at that point is the exchange of things of value consummated. The ledger has three important properties: replication, which synchronizes all of the copies of the ledger in the network; consensus, which assures that all ledgers are exact copies; and permissions, which ensure that members of a network can only see items in ledger that involve them.

When an entry is agreed to and committed to the blockchain's shared ledger, it cannot be changed. This is a critical feature, which differentiates blockchain's ledger from most database technologies—where entries can be updated and deleted. This makes blockchain resistant to tampering and provides clear audit trails for parties in transactions and government investigators to follow.

Another critical element of blockchain technology is the "smart contract." These are terms of agreement that are captured in software and stored and executed within the blockchain. The smart contracts automatically fulfill the obligations that members have agreed to. A blockchain is an ideal place to store and run such contracts because of its immutability and cryptographic security.

In our view, however, most blockchain implementations, and the tools surrounding them, aren't yet ready for many serious businesses use. The concept and architecture are taking form, but some key capabilities and standards are missing or only now emerging. For instance, many enterprise applications require more extensive security capabilities than most of today's blockchain implementations offer. Within healthcare, more extensive privacy protections are needed.

So, IBM and others in the industry are augmenting the core blockchain technologies with additional features. One goal is to ensure that institutions and individuals (whether participants or not) can only access information they're supposed to see. A key element is "entitled access," which is achieved by using modern cryptography so access to private data requires presentation of encryption keys/certificates held by authorized participants.

We're also taking steps to ensure that participants cannot commit fraud or collude in ways that jeopardize the integrity of the blockchain. Fraud and collusion resistance is achieved by ensuring that every transaction is validated by all the members of the blockchain networks, which might include regulatory and clearinghouse institutions.

Lastly, we're enabling regulators, with permission, to check for regulatory compliance, and for law enforcement with proper judicial authority, to access details of transactions in the course of criminal investigations.

These additional features will be essential in healthcare scenarios, where the privacy of individuals is both a legal and moral imperative. Blockchain can prevent against accidental or malicious privacy breaches by requiring both encryption and multiple signatures to approve access to sensitive information. There might be a mechanism, for instance, that for a patient record to be seen, a doctor, a nurse and the patient must approve within the blockchain.

Part 3: Why it's Critical for Blockchains to be Open and Interoperable

It's essential for blockchain technology to be developed following the open source model so a critical mass of organizations will coalesce around it—and reap its full benefits. Because of open source rules, participants can trust that the technology will fulfill their needs and conform with industry standards—assuring interoperability between blockchain applications. Also, by sharing the foundational layer, the participants can focus their individual efforts on industry-specific applications, platforms, and hardware systems to support transactions.

An open source blockchain with liberal licensing terms and strict governance will enable the broadest adoption of blockchain by regulated industries. The liberal licensing terms will accelerate innovation, and the strict governance will hasten adoption and regulatory acceptance.

Given the nature of a blockchain network, industry users and regulators of blockchain are going to want visibility right down to the source code to verify its source, accuracy and security.

We believe that the best path forward for blockchain is for the tech industry, government, and the business community to consolidate their efforts around a single open source blockchain foundation that's developed and governed in an environment of transparency and cooperation. We also believe that organizations will be best served if they use industry-specific or function-specific extensions of that technology, which are created and governed following the same principles. An example of this might be a banking framework that deals with loans, lenders, and borrowers.

There are several open source blockchain projects, but only the project managed and sanctioned by the Linux Foundation, called Linux Hyperledger, offers industry-friendly terms and multi-company governance. That's why we're participating in the Linux Hyperledger project and urging others to do so as well.

The Linux Foundation announced the project last December. Founding members of the initiative represent a diverse group of stakeholders, including ABN AMRO, Accenture, ANZ Bank, BNY Mellon, Cisco, The Depository Trust & Clearing Corporation (DTCC), Deutsche Börse Group, Digital Asset Holdings, Fujitsu Limited, IBM, Intel, J.P. Morgan, R3, Red Hat, SWIFT, VMware, and Wells Fargo. Already, several companies, including IBM, have contributed high-quality software code, technology, and intellectual property rights. The transparency, collaboration and shared governance of this project makes it attractive to participants—whether they're technology companies or enterprises who want to deploy the technology. The reaction to the announcement was overwhelming. More than 2300 organizations or individuals have asked to participate, the highest such tally in the Linux Foundation's history.

Part 4: Government's Stake in Blockchain

Blockchain is a true technology phenomenon. Less than a year ago, it was little known outside a small group of technologists. Now, it's making headlines everywhere and businesses and governments are scrambling to come to terms with it.

The good news for government leaders is that blockchain has the potential to transform governmental processes as fundamentally as it does those of the businesses—providing superior levels of transparency, accuracy and efficiency. It could help governments do everything from collect taxes and deliver social services benefits, to manage land registries, and assure the integrity of government records.

Take the US Social Security system, for instance. It involves the federal government, millions of employers, their payroll service providers, and more than 200 million beneficiaries and working individuals who are paying into the system. This is a model scenario for blockchain. There are many parties, many rules, many steps in the process of administering the system, and a critical need for very high levels of privacy protection and security from breaches.

Other potential uses of the technology are quite intriguing. What if the US government began issuing regulations and monitoring compliance via blockchain technology? And what if the government implemented the taxation system with blockchain? Individuals and businesses might never have to file an income tax return. Instead, a blockchain network noting their tax obligations and recording their financial transactions would continuously invoke the tax code, assess taxes and transfer money. No need to file a tax return.

Recommendations for Congress and Obama Administration

The possibilities are endless, yet most governments around the world have not yet begun to come to terms with blockchain.

In my view, there's a clear role for government—cribbed liberally from a position paper issued recently by the UK government. It should:

Use blockchain technology. Government should act as an early adopter and start deploying the technology for projects like voting, recording land registries, managing immigration, and the like.

Invest in research. Just as the National Institute of Standards and Technology works with industry to develop and apply technology, measurements, and standards, the government should investigate to make sure blockchain technology is robust, secure and scalable, while understanding the ethical and social implications of potential uses and the costs and benefits of adoption.

Create a regulation framework. The government needs to make sure that blockchains are being used in accordance with US laws while avoiding the stifling of innovation through excessive or rigid regulations.

Set standards to ensure security and privacy. The government needs to work with academia and industry to ensure that standards are set for the integrity, security and privacy of distributed ledgers and their contents. These standards need to be reflected in both regulatory and software code.

Conclusion

Blockchain is a classic emergent technology. It appears to have a broad set of uses and benefits, but it's so strikingly different from what people are used to that many business and government leaders alike are adopting a wait-and-watch attitude. We applaud judicious caution, but at the same time, we believe that organizations and institutions that don't quickly assess the potential of blockchain and begin experimenting with it risk falling behind as the world undergoes what we see as a tectonic shift.

Therefore, we urge Congress and the Obama administration to study and discover the best uses of blockchain for the US government and the best regulatory approaches to maximizing its potential while protecting the interests of citizens. Blockchain may have begun its existence in the shadows of the crypto currency realm, but it now stands in the open—a powerful tool ready to serve business and society.

TESTIMONY 2 – FEBUARY 9, 2016

Commission on Enhancing National Cybersecurity
Established by Executive Order 13718
New York University School of Law – Vanderbilt Hall
Jerry Cuomo, IBM Fellow, VP Blockchain Technologies

Eighty years ago, IBM assisted the U.S. government with creating the Social Security system. At the time, it was considered the most complex system ever developed. As financial systems are increasingly digital and networked, the public and private sectors again need to combine forces to make the financial systems of the future more efficient, effective and secure than those of the past. Social security numbers became the key to personal identity for an entire generation of systems. Today, institutions must collaborate to create digital methods for establishing identity to secure a new generation of transactional systems.

Blockchain technology is becoming an essential tool as businesses and society navigate the shift. It holds the potential to transform commerce and the interactions between governments and individuals. It provides trust and security, but for it to reach its full potential, it must be further developed and rely on open source to make it deployable on a grand scale. There is a critical role for government to enhance national security and competitiveness. The government must invest in scientific research to accelerate progress.

NIST can help set standards for interoperability, privacy, and security. Government agencies can become early adopters of blockchain applications. The government can certify the identities of participants in a blockchain-based system. Blockchain came to prominence as a core technology underlying Bitcoin. Industries and government agencies are now exploring the use of blockchain where transaction participants are known.

This is known as "permissioned blockchain." A distributed ledger shared on a peer-to-peer network contains an ever-expanding list of data records. Each participant has an exact copy of the ledger data. Additions to the chain are propagated in real time across the network with all parties consenting on the validity of the entries. Everyone in the chain has an up-to-date ledger with the most recent transactions. Once transactions appear in the ledger, they cannot be changed. Cryptography and digital signatures are used to verify identity authenticity and enforce access to the ledger.

It is fast and resolves transactions immediately, eliminates cost, and is resistant to tampering, collusion, and cyber threats. IBM is a founding member of the Linux Foundation's Open Source Hyperledger Project, helping to build foundational elements of a business-ready blockchain with privacy, confidentiality, and auditability. IBM is pioneering the use of blockchain to reengineer some of IBM's business processes.

There are four areas to enhance:

Proof of Identity – The Social Security Number (SSN) is not a secure or certifiable enough identity tool for a blockchain-based ecosystem. A new identity management system must be created. As an example, India is issuing 12-digit identity numbers to its entire population. An individual's number is linked to biometric and demographic information and can be used to set up a bank account or access government services.

Data provenance – This ensures the safety of exposing personal and confidential data to financial applications. New systems must track every change to financial data so that it is auditable and completely trustworthy. Data provenance must provide fingerprints and time stamps to show that information is up-to-date. It must be accurate, up to date and un-tampered with.

Secure transaction processing – While the parties in the transaction using blockchain are known to the system, the individuals only have access to the details of the transaction. The entities validate transactions to verify the contracts are being fulfilled without revealing confidential information to them. This involves homomorphic encryption. It makes it possible to verify information without having to de-encrypt it. This technique is still several years from being practical, but it is coming.

Sharing intelligence – The good guys are under pressure to change the game due to a rising tide of threats and cyber terrorism. Blockchain has that game-changing potential. It is more secure than other networks, it can be used to share threat information. Many financial services firms are reluctant to share threat information, but blockchain makes it possible to share securely. Threats can be shared in real-time so that accounting measures can be taken. IBM looks forward to working with government and industry to get this done.

TESTIMONY 3 – FEBRUARY 14, 2018

House Committee on Science, Space and Technology
Subcommittee on Oversight & Subcommittee Research and Technology
"Beyond Bitcoin: Emerging Applications for Blockchain Technology"
Gennaro (Jerry) Cuomo, IBM Fellow

Introduction
Good morning, Chairman Abraham, Chairwoman Comstock, Ranking Member Beyer, Ranking Member Lipinski and Members of the Subcommittees.

My name is Jerry Cuomo, and I'm the Vice President for Blockchain Technologies, at IBM.

Thank you very much for the opportunity to testify this morning.

We at IBM believe that blockchain is a revolutionary technology. With blockchain, we can reimagine many of the world's most fundamental business processes and open the door to new styles of digital interactions that we have yet to imagine.

You are wise to explore the science of blockchain technology—and its potential applications beyond cryptocurrency and financial technology—because blockchain has the potential to vastly reduce the cost and complexity of getting things done across industries and government.

Today, my testimony will share some key beliefs we hold at IBM based on our experience as an industry leader in blockchain. I'll also share some concrete examples that illustrate the transformative power of blockchain. Finally, I will include some recommendations for Congress and the Trump Administration that could ultimately help U.S. competitiveness and our citizens by preparing, advancing and applying blockchain in new ways—as I believe we should.

IBM's Blockchain Beliefs

Most people who have heard of blockchain associate it with the cryptocurrency Bitcoin. While they are related, it is important to understand they are not the same thing. Bitcoin is merely one example of a use of blockchain technology. Whereas Bitcoin operates with a network of anonymous participants, blockchain can also be used as a trusted network, using permissioning, to handle interactions between known parties. As an analogy, the internet, like blockchain, is a transformational building block for many types of communication, Bitcoin and other forms of cryptocurrency are but one use of blockchain, just as social media is but one use of the internet.

We have worked with clients in over 400 blockchain projects across supply chain, financial services, government, healthcare, travel and transportation, insurance, chemicals and petroleum, and more.

This experience has led us to develop three key beliefs that I'd like to share with you today:

1. Blockchain is a transformative technology.

2. Blockchain must be open.

3.   Blockchain is ready for business and government use TODAY.

Blockchain Belief #1 – Blockchain is a transformative technology
First and foremost, blockchain is changing the game. In today's digitally networked world, no single institution works in isolation.

At the center of a blockchain is this notion of a shared immutable ledger. You see, members of a blockchain network each have an exact copy of the ledger. New entries in the ledger are propagated throughout the network. Therefore, all participants in an interaction have an up-to-date ledger that reflects the most recent transactions and these transactions, once entered, cannot be changed on the ledger.

Blockchain's power to transform is that it enables co-development of a shared copy of the truth. And with this, **what a group can achieve together far exceeds what any individual member can achieve by themselves.**

Now let me tell you how blockchain actually changes the game.

- **Time** is saved because multi-party transactions can settle immediately avoiding exhaustive reconciliation that often take days or even months.

- **Cost** is reduced because business-to-business processing eliminates overhead caused by "middle-men."

- **Risk** is mitigated because the ledger acts as an immutable audit trail greatly reducing the chances for tampering and collusion.

This leads to my first example, IBM and Maersk, the world's largest shipping company, recently announced their intention to form a joint venture to create an industry-wide trading platform for the ocean freight industry. This industry accounts for 90 percent of goods shipped in global trade. Currently, one shipment of goods between two ports can generate a sea of paper and information exchanges between 30 different public and private organizations. The joint venture will use blockchain to help track in real-time millions of shipping containers across the world by providing a trusted, tamper-proof, cross-border system for digitized trade documents. By having a shared blockchain ledger, companies can reduce the time spent resolving disputes, finding information, and verifying transactions, leading to quicker settlement. When adopted at scale, the solution has the potential to save billions of dollars. This is the transformative power of blockchain applied to the shipping industry.

And blockchain technology provides the springboard for an even broader spectrum of innovation. Let me just take a moment to tell you about a project from the IBM research lab. Uniquely identifying a physical asset such as a type of a diamond, petroleum, or a manufactured part as a corresponding digital asset in a blockchain network is an interesting challenge; verifying authenticity is important.

These physical products travel through many hands and companies before reaching their final destinations. At any point along the supply chain, a valuable physical asset could have been swapped with a counterfeit one. To help ensure provenance on the blockchain, at IBM Research, we invented a smartphone-based Artificial Intelligence Technology used to scan the high value item. Using light spectral analysis to capture the microscopic properties, viscosity and other identifiers, the technology creates a digital fingerprint that can be used to verify authenticity and avoid counterfeiting documents or fake substitute products.

Blockchain Belief #2 – Blockchain must be open
For blockchain to fulfill its potential, it must be based on non-proprietary technology… Doing so will encourage broad adoption… and ensure the compatibility and interoperability of systems. Specifically, this enterprise-ready blockchain must be built using open source software, with a combination of flexible licensing terms and strict governance by an open community, meaning there is no one controlling organization that governs

the direction of the project and no lock-in to one vendor. Much as we have seen with the internet, only with openness will blockchain be widely adopted and enable innovation.

For this reason, IBM is participating with over 180 industry players in the Hyperledger organization, led by the Linux Foundation. Hyperledger is a collaborative open-source, open-standards and open-governance effort created to advance cross-industry blockchain technologies for business and government.

For example, IBM is collaborating with companies like SecureKey and the Sovrin Foundation on blockchain-based digital identity. Together, we are working to create a global ecosystem of blockchain identity networks backed by global standards. These standards are defining mechanisms by which only the information that needs to be shared is shared with only those parties that need to know. With blockchain, identity theft and fraud can be significantly reduced while at the same time increasing the effectiveness of Know-Your-Customer and Anti-Money Laundering efforts, doing so in a more cost-effective way. We can not only make it harder for criminals to impersonate someone, but in the event of a data breach, we can recover quickly. Unlike a social security number, blockchain backed decentralized identifiers can easily be revoked and reissued if ever stolen or compromised.

Blockchain Belief #3 –Blockchain is ready for business and government use TODAY

Not all blockchain technology is created equal. For broad business and government use, enterprise blockchain technology is now available that solves four fundamental requirements: accountability, privacy, scalability and security.

Accountability means the participants transacting in a network, and the data they are transacting on, are both known and trusted. In an enterprise ready blockchain, participants are known and are identified by membership keys. The data can be trusted because transactions committed to the ledger are immutable—such that they cannot be removed or changed by the actions of a single party.

With this accountability, the network is auditable allowing members to follow and adhere to existing government regulations like HIPAA and GDPR.

Even though participants are known, they must be able to transact with **privacy** on the network. Businesses require that both their transaction data and the transactions themselves are confidential. An enterprise blockchain enables confidential communications when information is not desired to be shared with the entire network.

Computer systems and networks must be architected to have the **scalability** to handle an immense volume of transactions. Because trust in an enterprise blockchain network is not built through anonymous "mining" (as is done in Bitcoin), transaction performance has been demonstrated at levels needed for high volume throughput. A recently published research paper demonstrated one such enterprise blockchain performance at a best-of-class rate of more than 3500 transactions per second (https://arxiv.org/abs/1801.10228).

The need for **security** continues to be illuminated by breaches in the news every month. As much as everyone tries, it's impossible to eliminate all people with malicious intent or sloppy actions. An enterprise blockchain network is fault tolerant, implementing algorithms like crash and Byzantine Fault Tolerance that allow a network to continue to operate even in the presence of bad actors or carelessness.

These four requirements are delivered today in the Hyperledger Fabric, one of the popular blockchain frameworks from the Hyperledger project. It now serves as the basis for over 40 active blockchain networks that are running on the IBM Blockchain Platform.

For example, every year 400,000 people around the world die from foodborne illness. With the advent of global supply chains, it's very difficult to trace contaminated food back to the source, as we witnessed with the recent e. Coli outbreak that sickened 60 people and took 2 lives over a period 6 weeks. IBM is working with twelve major food companies—including Walmart, Unilever, and Nestle—applying our enterprise blockchain to rapidly trace food as it moves from farm to table, showing how blockchain has the potential to help keep entire populations healthier. Blockchain makes it possible to *quickly* pinpoint the source of contamination, reduce the impact of food recalls, and limit the number of people who get sick or die from foodborne illness.

Recommendations for Congress and Trump Administration

We are working with many government entities on activities for the adoption and use of blockchain technology: from the Smart Dubai initiative to trusted digital identity projects in Canada. U.S. companies are leading in blockchain technology development while U.S. government agencies like NIST actively engage in blockchain standards exploration. IBM and the FDA are exploring how blockchain can provide benefits to public health by reducing complexity in interactions between clinical trial sponsors, investigators, CROs and patients. While blockchain remains a team sport, there is an opportunity for the United States to build upon its momentum to lead blockchain by doing. I'd like to make a few recommendations to help Congress and the Trump Administration along this path.

First, we should focus our efforts on projects that can positively impact U.S. economic competitiveness, citizens, and businesses. The Congressional Blockchain Caucus, led by Reps. Jared Polis and David Schweikert, has already begun this critical work. The Blockchain Caucus is working to collect information on blockchain projects that could help individuals securely establish their identity, make key payments—such as tax payments—and revolutionize supply chains. This work should fuel initiatives that can make a meaningful difference in citizens' lives like the digital identity, food safety, and transport supply chain examples we discussed as well as other potential use cases like land registration, taxation, and more. I recommend we explore blockchain adoption and use with these citizen and business-focused projects first. Then, use the knowledge gained to inform policy and regulation in different blockchain technology implementations going forward.

Second, thoughtfully inserting blockchain in appropriate projects already funded would help ensure we stay at the forefront of this transformative technology. Consider blockchain technology as an enabler to lower cost, time and risk in currently budgeted projects. In addition, look for opportunities to fuel innovation in the broader ecosystem of U.S. businesses by encouraging blockchain projects as part of the Small Business Innovation Research (SBIR) program that is part of existing research budgets in a number of agencies today.

Finally, and perhaps most importantly, recognize the difference between blockchain's use in new forms of currency from broader uses of blockchain when considering regulatory policy. Carefully evaluate policies established regarding cryptocurrencies to ensure that there will not be unintended consequences that stymie the innovation and development surrounding blockchain. A policy that has not been carefully vetted could risk inhibiting the U.S. leadership position.

Blockchain is ready for the government, now let's get the government ready for blockchain. We at IBM stand ready to help provide the analysis of any such policies and would be happy to work collaboratively with Congress to ensure the continued expansion and success of blockchain.

Conclusion

Thank you for the opportunity to discuss such an important topic for our present and our future. In summary, at IBM we believe that:

- Blockchain is a transformative technology: It enables many to achieve more than what is possible by one.

- Blockchain must be open: Only then will blockchain be widely adopted as a springboard for innovation.

  - Blockchain is ready for business and government use TODAY: it provides accountability, privacy, scalability, and security.

At IBM, we are actively working to ensure businesses and governments are thoughtfully implementing this technology and we believe the United States has an opportunity to lead in this space.

I will look forward to answering your questions and continuing this discussion. Thank you.

UP NEXT

Yes, of course, my camera roll has several bow-tie filled photos of my testimonies to US government. Here are a couple from the 2016 and 2018 sessions.

Addendum 1 - Figure 1 – Jerry's testifying to US Congress in 2016 and 2018

Thankfully, I had plenty of help preparing and planning for testimonies. I am glad the camera caught Meeta Vouk (below, left side), Mark O'Riley (center) and Peter Rieth (right side), each of whom played a vital role in ensuring I didn't freeze during my speeches to Congress.

Addendum 1 - Figure 2 – Jerry with colleagues "on the hill".

- - -

One more supplemental chapter. I think you will enjoy this one.

Addendum 2 -
I'M SATOSHI?

This supplemental chapter was written by my friend and colleague Mark Parzygnat. The idea for this chapter was tossed around during a conversation between Mark and me, as we reflected that "not knowing who invented Bitcoin" was a big deal, and unprecedented in modern times. When you think about it, we pretty much know who contributed to the introduction of all major advancements in technology. Mark observed that perhaps not since the invention of fire or the wheel can we not track the ownership to a group or individual. In the chapter, a profile of Satoshi Nakamoto is established that looks and tries to match it against a few known candidates. Mark also boldly identifies who he thinks the real Satoshi is. Will you agree? The chapter takes a lighthearted approach to answering the question of who Satoshi is, while also looking at the important phenomenon that launched decentralized cryptocurrency and got us all to "*Think Blockchain*".

WHY THIS CHAPTER?

Bitcoin was built using blockchain technology because of the philosophy of being open and transparent. To the contrary this is completely opposite of the creator. Blockchain and their applications are always noted for the ability to trace back to the origin, or to the genesis block. Blockchain's fame began to be realized when Bitcoin entered the scene, endurably linking them together.

Addendum 2 – Figure 1 – Satoshi is the "missing genesis block". A woman? A man? A group?

If Bitcoin's history were itself a blockchain, it would be missing the genesis block. We would see the previous block of **cryptocurrency** linked to Bitcoin. **Bitcoin's** previous block would be linked to the first application of **blockchain technology**, and previous block of this clever use of blockchain technology (with cryptography, distributed computing and gamified mining) would be linked to "null." The genesis block is missing. So, who is the actual true inventor and what makes them tick?

In this supplemental chapter, we will take a look at what we do know about the inventor of Bitcoin, along with reasons it could be important to keep Satoshi anonymous. We will dive into what knowledge and ability it takes to conceive something as complex as Bitcoin, candidates that have the potential to be Satoshi, and why it was created.

With that, let's get going. It's normal and natural when someone begins learning or digging into blockchain as a technology, it quickly gets linked to Bitcoin. Again, while Bitcoin has accelerated blockchain into the limelight, Bitcoin and blockchain are not one and the same. Particularly in Bitcoin's case, the story behind its creator is very unusual, intriguing, and worth a little time of investigation. This allows for background on the reasons, theories, and the direction this particular area of interest was created and the direction it was intended to be driven in.

We do know that the inventor is identified as Satoshi Nakamoto. However, we don't know the exact person behind the name itself. There are not many significant accomplishments outside of the wheel and fire that have had such huge exposure, where the lineage can't be linked back to the inventor. While Bitcoin may not impact every individual's daily life like fire or the wheel, it is quite a profound innovation, and similarly, the inventor (or inventors) is to this date still unknown. All other cryptocurrency or altcoins have known inventors. The mystery and intrigue, however, envelop many in the IT industry.

PROFILE OF SATOSHI NAKAMOTO

When I first began in blockchain and started learning about Satoshi, I became infatuated with the vast body of knowledge it would take to create Bitcoin. Sit back for a minute and think about just some of the obvious factors. First to collate and understand all the information that is needed regarding the current state of financial systems, fiat, and the potential inherent issues involved. Some of these issues include a central authority, loans without backing, inability to make micro payments, and double spend, as discussed previously in this book. Next, the ability to completely comprehend the nuances of these problems. With such an understanding to be able to research and discover a method which he completely believed in that could overcome these problems. Next, add in the required knowledge to deliver all these pieces in complex computer-science algorithms on top of a very nascent technology of blockchain.

Of course, we know that he didn't deliver everything that we know as Bitcoin today, without help from others. Which adds in an additional element of the persona— the social and communal aspect. While Satoshi may have delivered the paper and some of the first iterations of Bitcoin, he also built a community behind it, and quickly. The profile requires being a competent and trusted entity with the ability to create a following, a philosophy... Heck a movement! This movement delivered the cryptocurrency in a matter of months and continues to thrive and grow today. Seems very difficult to believe a single person could achieve this monumental task solo. To contradict all this a little, consider the fact that to date nobody truly knows, or at least has come forward in an undeniable way to prove they are, in fact, Satoshi Nakamoto. Bitcoin being built on top of blockchain introduces traceability as we have previously discussed. This means that the real genesis block of Bitcoin can be traced back and that Satoshi's "address" could be linked to many of the first transactions. In fact, at prices cited in November of 2021, it's valued at over $56 billion. Excellence in computer engineering and financial services can sometimes bring along egos. So, how are these proud creators keeping this a complete secret, and someone not coming forward to cash in? And secrets are exponentially more difficult to keep when you add more personalities to the group.

To claim or not to claim, that is the question.
Why would someone not want to step forward and claim the prize?

Well, if claiming means selling, then perhaps it's because of fear of cratering Bitcoin. Think about it. A mass influx of over a million bitcoins would cause the market to crash. There is a capped volume of bitcoin allowed to be created, the difficulty of possessing a limited resource is what drives price. Then the social ramification of seeing the founder and creator of the currency appearing to have lost faith in his creation and doing a "dump and run" would turn the community into a frenzy of fear selling, completely devastating the currency.

Does knowing Satoshi centralize Bitcoin?

Satoshi had quite an influence on the code base and direction early on. Perhaps this can be likened to the influence that Vitalik Buterin has over Ethereum today. Satoshi might catch similar criticism from skeptics saying that when things go wrong with the decentralized Ethereum network, Vitalik comes to the rescue by centralizing it for a moment while he fixes the problem. Hence, for Satoshi to suddenly be known, and or be as active as he used to be in the community, could result in unintended influence and decisions in a hierarchical or ownership sense.

WHO IS SATOSHI? LET'S TAKE SOME GUESSES

Now that we have taken into account factors that build out the persona's profile, who is Satoshi Nakamoto? This is a question that has been asked millions of times, and I wish I could tell you for sure. It is common belief that the name is a pseudonym, used to protect the identity.

Some would state that the NSA has uncovered the identity, using a series of techniques matching common language and terms from known posts to create a "fingerprint," then using that fingerprint to match to a specific individual, through social media and common search engines. Much of the world would still claim Satoshi's identity as unknown. Again, the simple fact is, to date nobody has come forward to prove without a shadow of doubt that they are Satoshi Nakamoto.

Another level of complexity for identification is common belief that it is not just a single individual but rather a group of folks. The degree of complexity and diversity of skills to create Bitcoin is just one reason to suspect that Satoshi represents a group. The one thing for sure is that Bitcoin was created to not have a single source with ownership or authority of controlling all aspects. This is another reason why it's important that the true-identity stays unknown. The influence this person or persons could have over decisions made has the potential of being immense, making it the clear and obvious choice for the development of the cryptocurrency is via an open-source community.

With that said, there have been several individuals thought to be or claiming to be the entity behind the name. Here are just some of the folks, and the reasons behind the theories.

The most likely contenders

Dorian Nakamoto – A computer engineer who was in 2014, presented by *Newsweek* reporter Leah McGrath Goodman as Satoshi due to similar political views, temperament, and mathematical skill. Dorian was interviewed on what he thought was regarding computer engineering, and not Bitcoin. When the conversation steered toward questions asking if he was Satoshi Nakamoto, he adamantly denied being Satoshi.[73] Satoshi published an online statement in a Bitcoin forum, confirming Dorian was not in fact the creator of Bitcoin.[74]

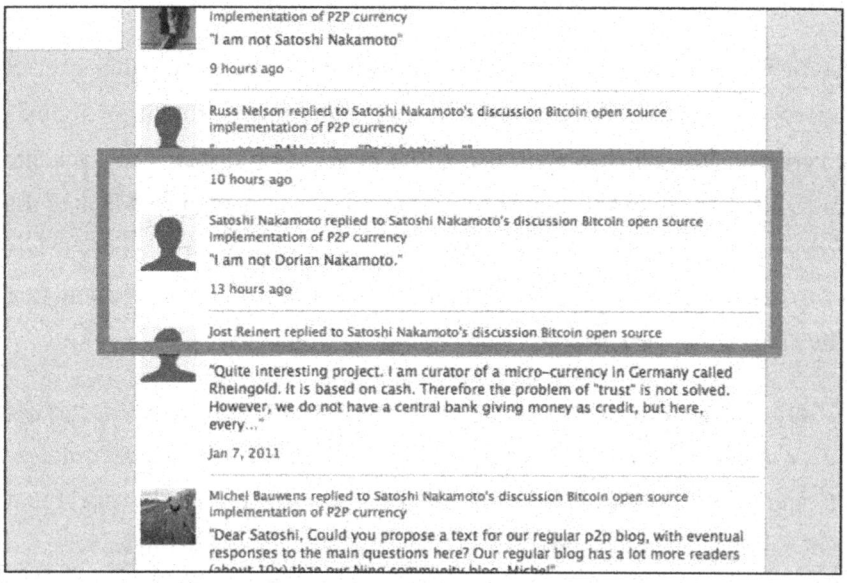

Addendum 2 – Figure 2 – Social media posts denying Satoshi's identity

Nick Szabo – He is the Founder of Bit Gold, with a background in digital contracts and digital currency, acquired from his computer science and law background. Nick published papers in the 1990s about Bit Gold, as a likely precursor, well before the Bitcoin's emergence. In 2014, he even tweeted "Not Satoshi – but thank you".[75]

However, in 2014, The Aston University Centre for Forensic Linguistics conducted a linguistic study on Bitcoin's white paper to determine the identity of Nakamoto. The group concluded that Nick Szabo was Nakamoto based on linguistic similarities.

While it is not clear that Nick is "the one and only Satoshi," if I were to bet, I would say Nick may have had a hand in it. He just matches the profile so well.

Hal Finney – A developer for PGP corporation and an early on contributor to Bitcoin, Hal received the first Bitcoin transaction from creator Satoshi Nakamoto. Hal published an autobiography titled *Bitcoin and Me*. Hal has always been a futurist, and had his body cryonically preserved after passing from ALS in 2014.

Again, like Nick, Hal could very well have had a hand in being the genesis block. Finney was famously discovered to have lived just blocks away from a Dorian Nakamoto in small town Temple City, CA.[76] Pretty interesting… but keep reading. It keeps getting better.

Len Sassaman – A highly skilled cryptographer and protocol developer, who was a huge advocate for privacy, providing him with the ability to build the Bitcoin foundation. He also worked with Hal Finney. Len and Hal had both worked on anonymous remailing technology.[77] Len committed suicide in 2011, just two months after Satoshi's last public correspondence stating he would probably not be around in the future. Sassaman has a tribute that is embedded into Bitcoin.[78]

Others to consider
Elon Musk – His innate ability to innovate across fields has made him a possible candidate. Elon has been able to create innovations such as Hyper-

loop and Space X from researching and reading material on subjects. He has repeatedly denied being the creator of Bitcoin.

Craig Wright – Several have self-proclaimed to be the creator of Bitcoin, one of the most famous being Craig Wright. In 2016, Wright's proclamation resulted in a lawsuit from relatives of Craig's colleague, Dave Kleiman. Their claim being that Kleiman and Wright worked together during the period of time leading up to the release of Bitcoin. The relatives argued that given the close alignment of their collaboration timeline implies that Kleiman helped Wright invent the cryptocurrency. Therefore, their claim was to the Wight's estate entitling them to half of Satoshi's bitcoin stash equaling over $50 Billion.

The trial ran for several years from 2018-2021 with the jury not being able to prove or disprove Wright is in fact Satoshi Nakamoto. One of the biggest defense points was the fact that Wright has never moved any of the original bitcoin from Satoshi's stash.

DRUM ROLL PLEASE...

Well, with the contestants enumerated above, I am ready to unveil, in my humble opinion, who Satoshi is. First and foremost, for the reasons cited above, I believe Satoshi is a group of people. Specifically, I believe the Satoshi Nakamoto pseudonym was created by a collaboration between:

Dorian Nakamoto, Nick Szabo, Hal Finney, and Len Sassaman

While that's where the evidence seems to point, there is no way to prove me right or wrong.

At this point in time, we simply don't know who the real Satoshi is. It could be someone you walk by on the street or meet at a blockchain convention. They might even be admitting their identity in plain sight.

Addendum 2 – Figure 3 – Chris Ferris at a blockchain
convention wearing an "I am Satoshi" t-shirt

Looking a little deeper into the pseudonym itself. Why was this name even chosen? Looking at the meaning behind names, it's interesting to find, the name 'Satoshi' is said to mean or translate to "wise and clear thinking." 'Nakamoto' translates to "one that lives in the middle", or "of central origin," which is fitting in respect to the purpose of Bitcoin and the removal of the big financial institutions as the middleman of financial transactions. Some say, this is what was responsible for the financial collapse of 2008.

However, Satoshi Nakamoto didn't come into view just from Bitcoin's rise to fame. He was already a familiar name or user identity prior to Bitcoin's appearance. Years prior, there were many correspondences and interactions with cryptographic developers via online message boards and emails under the same name.

MISSION AND REASON BEHIND HIS WORK

With the background on Satoshi, the question that comes up naturally is: Why did he create Bitcoin? This is a simpler question with a definitive answer. His mission was clear. To take control back from untrusted financial intermediaries through a decentralized financial system. The 2008 subprime mortgage crisis as a result of a housing market collapse, showed how fragile the state of the financial system truly was.

Bitcoin, on the other hand, allowed for individuals to have a system of trust through transparency among participants. Ultimately providing an alternate peer-to-peer payment system free of centralized controls with lower costs. A copy of his initial letter detailing why and what he had developed can be found here on the Bitcoin forum. He also referenced a white paper that he published on a cryptography mailing list entitled "Bitcoin: A Peer-Peer Electronic Cash System". The paper, published on October 31, 2008, outlined a decentralized peer-to-peer protocol that was cryptographically secure. This was just three months prior to the mining of the first Bitcoin. The first Bitcoin was mined on January 3, 2009, known as the genesis block.

WHERE IS SATOSHI NOW?

Who knows, maybe he is out in the world teaching and sharing experiences gained through blockchain-related world travel. Just like these two guys.

Addendum 2 – Figure 4 – Jerry and Mark visiting the Great Wall of China after presenting together at a conference in Beijing, China

It's another unknown, like his identity. What we do know is on April 23, 2011, he backed away from the development of Bitcoin and the cryptocurrency scene. This happened like any bad breakup, via an email to a fellow developer stating that he is moving on to other things and the cryptocurrency future is "in good hands." Really sounds like an "It's not you, it's me" kind of text message of today. If the reason for the departure from the Bitcoin scene was known, it might lead to the identity. The sudden exit though, doesn't stop one from speculating about the reason.

Maybe, the disappearance was a result of WikiLeaks' plan to use Bitcoin as its financial backing. Wikileaks was being blocked from all other major payment processors. They were blacklisted by the U.S. and Australian governments after obtaining thousands of classified documents regarding the war in Afghanistan and releasing them to several news outlets.

On December 5th, 2010, Satoshi made an appeal to WikiLeaks to not consider Bitcoin, stating, "I make this appeal to WikiLeaks not to try to use Bitcoin. Bitcoin is a small beta community in its infancy. You would not stand to get more than pocket change, and the heat you would bring would likely destroy us at this stage." A week later was Satoshi's last public statement. Satoshi described this as a "kick to the hornet's nest" and was very much against the idea as it would bring with it a lot of unwanted government attention, which was bound to happen if Satoshi's creation became successful at any point. Other theories are tied closer to individuals discussed in this chapter and that ailing health physically or mentally may have led to the decision to remove himself prior to passing.

HATS OFF TO SATOSHI!

While we don't have all the answers to the questions brought up in this chapter, at least we have a glimpse of the qualities and attributes of a person or team that created such an incredible technological accomplishment that continues to thrive today. Now ask yourself, if and when you come up with your world impacting idea, will you be able to deliver it by yourself, keep it completely anonymous, and watch it grow for decades? Especially knowing there is a fortune to be had that is just a simple transaction away? I doff my hat to Satoshi, as I don't think I could resist cashing out!

What have we learned in this section? First, there are a lot of questions still out there around the creator of Bitcoin. Who are they, where did they go, are they still alive, is it a single person or a team of folks? We may never know, but I have my suspicions that it's the group of four mentioned above. However, what we have learned is the reason that Bitcoin was created, which ultimately has led to thousands of follow-on cryptocurrencies competing in a different kind of, very volatile market. We have an understanding of the intellect and qualities it takes to come up with such a creation. We also learned there are a lot of folks that have been believed to possess these qualities and suggested to be Satoshi.

QUIZ TIME

1 - Who is Satoshi Nakamoto? Of course, this was going to be asked!

    a.   Jerry Cuomo

    b.   The electric car guy

    c.   Unknown but likely Dorian Nakamoto, Nick Szabo, Hal Finney, and Len Sassaman

    d.   Mark Parzygnat (A shameless plug for myself, I guarantee I am not Satoshi)

2 - How has Satoshi kept Bitcoin from aspects of centralization and single control?

    a.   Created it via open source

    b.   Removed himself from the scene so as to not sway the future in a specific direction

    c.   The design of Bitcoin removes central authorities

    d.   All of the above

Acknowledgements

MEET THE CONTRIBUTORS

Foreword by Dr. Irving Wladawsky-Berger

I am thrilled and honored to have Iriving Wladawsky-Berger write the foreword for this book. I have always looked up to Irving as a role model for my career, so his foreword is a badge of honor that my book shall proudly wear. Thanks again, my friend. Looking forward to our future (Web3) adventures... Jerry

Photo – Foreword Author: Irving Wladawsky-Berger (left),
Irving visit Second-Life Metaverse in 2006 (right)

Irving is a Research Affiliate at MIT's Sloan School of Management and Cybersecurity at MIT Sloan (CAMS) and Fellow of the Initiative on the Digital Economy, of MIT Connection Science, and of the Stanford Digital Economy Lab.

He retired from IBM in May of 2007 after 37 years with the company, where he was responsible for identifying emerging technologies and marketplace developments that are critical to the future of the IT industry, initiatives including the Internet, supercomputing, and Linux.

Since retiring from IBM, Irving has been an Adviser on Digital Strategy and Innovation at Citigroup, at HBO, and at Mastercard. He's been writing a weekly blog, irvingwb.com, since 2005. From April of 2012 until July 2020, Irving was a guest columnist for the *Wall Street Journal's CIO Journal*.

Irving served on and later became co-chair of the President's Information Technology Advisory Committee from 1997 to 2001 and was a founding member of the Computer Sciences and Telecommunications Board of the National Research Council in 1986. Irving is also a Fellow of the American Academy of Arts and Sciences. Born in Cuba and having come to the US at the age of 15, he was named 2001 Hispanic Engineer of the Year.

Irving has an M.S. and a Ph. D. in physics from the University of Chicago.

Illustrations by Shaun Lynch
Shaun and his illustrations have a way of bringing out my personality and fueling my storytelling in a manner that no other can... Thanks for the cover art and the many illustrations over the years that I cut and pasted into this book. You've brought a heightened touch of creativity by animating life into our blockchain ducks... Jerry

Photo – Illustrator: Shaun Lynch and Jerry having coffee at Starbucks in Raleigh, NC

Shaun is an award-winning (Red Dot, A* Design) British designer based in Austin, Texas, working at IBM on Automation and Blockchain projects. His professional interests lie primarily in designing for enterprise technology experiences as well as utilizing and maintaining design systems. He's also a keen illustrator in his free time with a soft spot for dogs (and ducks ;-))

"I'm Satoshi?" by Mark Parzygnat

To my partner in crime, Mark Parz, thank you for always so enthusiastically supporting my crazy projects. Your feedback and energetic guidance are priceless. And I am so glad that you volunteered your "I'm Satoshi" piece for this book. I look forward to returning the favor by adding a piece in your future book… Jerry

[See photograph of Mark and Jerry in his supplemental chapter "I'm Satoshi?"]

Mark Parzygnat is the Director, Global Solution Architecture, for Red Hat. Mark joined Red Hat in 2022, after 24 years with IBM. Mark has held various positions including product development and support, product and release management, and continues to be a champion of emerging technology. He was one of the first involved in and helped create the blockchain division at IBM.

He also helped lead the effort to open source the Distributed Ledger Technology (DLT) known today as Hyperledger Fabric, through the Linux Foundation. Mark says his inspiration for coining the name Hyperledger "Fabric" came from his undergraduate degree in Textile Engineering from North Carolina State University. Mark is passionate about evangelizing technology that he believes in, such as blockchain, containers, and Kubernetes. He has many publications and patents in a variety of technology spaces. He likes to tinker with small engines and loves to spend time with his wife and kids.

Think Blockchain Labs by Barry Mosakowski

Yeah, Barry is a great friend and colleague, but the connection that defines us best is that we are the backline (as shown in the photo below) for the Mind the Gap rock band (Adolfo, Marc and Lin round out the band). Hence our greater calling in life is to keep the beat and rhythm going at a rock tempo, whether that be music or blockchain. Barry conceived of and operates "Think Blockchain Labs," which is the GitHub repo associated with his book. Thanks, Bearman!... Jerry

Photo – Think Blockchain Labs: Barry Mosakowski (drums) and Jerry (bass)

Barry Mosakowski is a solutions architect at Red Hat and a 25-year veteran of IBM. Barry has a background in quality, support, development, and technical sales. He led the quality team for the IBM Blockchain offerings and Hyperledger Fabric open source project as well as he served in a technical sales role and assisted in architecting blockchain solutions to customers. Prior to blockchain, Barry was the quality lead for the IBM DataPower Appliance and one of the creators of the DataPower virtual appliance. Barry has numerous patents in networking, security, and blockchain, and is currently on the CompTIA Blockchain Advisory Board.

Walker Mosakowski

I want to add a special thank you to Walker for his role as my smart contract and tokens consultant and for providing sample code and pointers for the Ethereum coding portions of this book. If anyone out there is looking for a rock-solid Solidity Developer, Walker is your guy!... Jerry

Blockchain for Business by Arun, Cuomo, Gaur

I also want to thank the co-authors of my first blockchain book, Blockchain for Business. *Working on the* Think Blockchain *project, gave me a much better appreciation for all the diligence and hard work that these guys put into the* Blockchain for Business *book to get it past the finish line. I found myself channeling their work ethic several times and must credit the influence that* Blockchain for Business *had on* Think Blockchain. *Thanks, guys. Looking forward to our next project together... Jerry*

Photo – Jai Arun and Jerry signing books at Think 2019 in San Francisco

The Art of Automation

If I'm not jamming with Mind the Gap, blockchain'ing or hanging out with my wife and children (and dogs), I am most likely hosting an episode of my podcast, The Art of Automation. *AI-powered automation is another technology area that I am quite passionate about. Inspired by advancements in technologies that led to self-driving cars, the podcast explores how similar technology can be used to create an auto-assist mode for your business. Following the podcast, I recently released* The Art of Automation, *a book that features technical deep* dives in key areas of automation (e.g., process mining, robotic process automation (RPA), and AI for IT Operations (AIOps)) and automation industry use cases (e.g., healthcare, retail, insurance). Royalties from the book go to the American Cancer Society. If you've enjoyed this book, you'll enjoy The Art of Automation, too. Check it out ... Jerry

Photo – The Art of Automation Book and Podcast on Spotify

Supporting Cast

Wait, how did these guys make it into the book? Easy answer. They are the essential members of my supporting cast. The cast pictured below are Stephanie (my wife), Sheldon, (in white), and Lenny (in brown). Thanks for not barking too much during my marathon writing weekends. And thanks for listening, and always sounding interested in the week-to-week updates on how the book is coming along. I draw my energy from your love... as well as the daily walks. (Bet you can't figure out which comments are for my wife, and which are for the dogs. Hint, my wife and dogs walk me, not the other way around ;--)) ... Jerry

Photo – My support cast

...

WELL... THAT'S ALL FOLKS...

References

Preface

1   Forget Bitcoin: Blockchain is the Future, Nathan Reiff , July 26, 2021, https://www.investopedia.com/tech/forget-bitcoin-blockchain-future/

Chapter 1 – Blockchain Ducks

2   Blockchain Cleveland, 2018 – Jerry Cuomo's talk https://twitter.com/BlocklandCLE/status/1069956428864069632

3   What Does ACID Mean in Database Systems? https://database.guide/what-is-acid-in-databases/

4   Distributed Ledger, Wikipedia https://en.wikipedia.org/wiki/Distributed_ ledger#cite_note-UKscienceOffice201601-1

5   The Blockchain Scalability Problem & the Race for Visa-Like Transaction Speed, Kenny L https://towardsdatascience.com/the-blockchain-scalability-problem-the-race-for-visa-like-transaction-speed-5cce48f9d44

6   The History of Bitcoin, Wikipedia https://en.wikipedia.org/wiki/History_of_bitcoin

7   34 Blockchain Applications and Real-World Use Cases Disrupting the Status Quo, Sam Daley https://builtin.com/blockchain/blockchain-applications

Chapter 2 – Three Blockchain Stories

8   Dole Spinach E. coli Outbreak, Dan Flynn, September 20, 2009 https://www.foodsafetynews.com/2009/09/ meaningful-outbreak-7-dole-spinach-e-coli-outbreak/

9   Why is it so difficult to trace the origins of food poisoning outbreaks?
    ScienceDaily, June 2012
    http://www.sciencedaily.com/releases/2012/06/120601103812.htm

10 Total Identity Fraud Losses Soar to $56 Billion in 2020, March 2021
    https://www.javelinstrategy.com/press-release/total-
    identity-fraud-losses-soar-56-billion-2020

11 Hacker Lexicon: What Are Zero-Knowledge Proofs? Lily Hay Newman,
    Sep 14, 2019 https://www.wired.com/story/zero-knowledge-proofs/

12 WHO urges governments to take action, November 2017
    https://www.who.int/news/item/28-11-2017-1-in-10-medical-
    products-in-developing-countries-is-substandard-or-falsified

13 Pairing AI with Optical Scanning for Real-World Product Authentication,
    Donna Dillenberger, May 23, 2018
    https://www.ibm.com/blogs/research/2018/05/ai-authentication-verifier/

Chapter 3 – How Blockchain Works

14 *How to time-stamp a digital document. Journal of Cryptology, Haber, Stuart;
    Stornetta, W. Scott, January 1991*
    CiteSeerX 10.1.1.46.8740. doi:10.1007/bf00196791. S2CID 14363020

15 Bitcoin: A Peer-to-Peer Electronic Cash System, Satoshi Nakamoto
    (satoshi@gmx.com), Oct. 31, 2008
    https://papers.ssrn.com/sol3/papers.cfm?abstract_id=3440802

16 An easter egg in the Bitcoin genesis block code, reddit.com
    https://www.reddit.com/r/Bitcoin/comments/9ni31e/
    an_easter_egg_in_the_bitcoin_genesis_block_code/

17 About Node.js, Nodejs.org
    https://nodejs.org/en/about/

18 What is npm? Nodes.org
    https://nodejs.org/en/knowledge/getting-started/npm/what-is-npm/

19 SHA-2, Wikipedia
    https://en.wikipedia.org/wiki/SHA-2

Chapter 4 – Cryptocurrency, Bitcoin and Mining

20 What is Cryptocurrency and how does it work?

https://www.kaspersky.com/resource-center/
definitions/what-is-cryptocurrency

21 Currency, Investopedia
https://www.investopedia.com/terms/c/currency.
asp current monetary system/

22 What is Bitcoin? Bitcoin Explained Simply for Dummies
https://www.youtube.com/watch?v=41JCpzvnn_0

23 Wells Fargo account fraud scandal, Wikipedia https://
en.wikipedia.org/wiki/Wells_Fargo_account_fraud_scandal

24 Emergency Economic Stabilization Act of 2008, Wikipedia https://
en.wikipedia.org/wiki/Emergency_Economic_Stabilization_Act_of_2008

25 2016 Indian banknote demonetization, Wikipedia
https://en.wikipedia.org/wiki/2016_Indian_banknote_demonetisation/

26 Public Key Infrastructure, Wikipedia
https://en.wikipedia.org/wiki/Public_key_infrastructure

27 What is Cryptocurrency Mining | Explained for Beginners
https://www.youtube.com/watch?v=2VtH-XAOjXw

28 How are Transactions Included in a Block?
https://medium.com/ethereum-grid/ethereum-101-how-
are-transactions-included-in-a-block-9ae5f491853f

29 What is an Altcoin?
https://bitflyer.com/en-eu/faq/55-7

30 Litecoin (LTC): The Silver to Bitcoin's Gold, Charlie Lee, October 21, 2021
https://www.gemini.com/cryptopedia/litecoin-vs-bitcoin-blockchain

Chapter 5 – Tokens, Ethereum and Smart Contracts

31 (Parts of this chapter were inspired by) A History of Tokens, Sean Coates,
Aug 15, 2018
https://medium.com/@fooyay/a-history-of-tokens-a064b28d5af2/

32 What types of digital assets exist?
https://www.galaxyfundmanagement.com/crypto-101/
what-types-of-digital-assets-exist

33 Jefferies Sees the NFT Market Reaching More Than $80B in Value by 2025,

Coindesk, Will Canny, Jan 20, 2022
https://www.coindesk.com/business/2022/01/20/jefferies-sees-the-nft-market-reaching-more-than-80-billion-in-value-by-2025/

34 Turing Completeness, Wikipedia
https://en.wikipedia.org/wiki/Turing_completeness

35 A Next-Generation Smart Contract and Decentralized Application Platform, Vitalik Buterin
https://ethereum.org/en/whitepaper/

36 What is a smart contract? Kelsey Ray, November 5, 2019
https://coincentral.com/what-is-a-smart-contract/

37 The Idea of Smart Contracts, 1997, Nick Szabo
https://www.fon.hum.uva.nl/rob/Courses/InformationInSpeech/CDROM/Literature/LOTwinterschool2006/szabo.best.vwh.net/idea.html/

38 Smart Contracts — What Are They and How Do They Work? Matt Stokes, Dec 6, 2021
https://medium.com/geekculture/smart-contracts-what-are-they-and-how-do-they-work-2a5de5ec4cab/

39 What Is Ethereum 2.0? Luke Conway, July 15, 2021 https://www.thestreet.com/crypto/ethereum/ethereum-2-upgrade-what-you-need-to-know/

40 Top Ethereum Alternatives and Competitors 2022: ETH Killers, Srujana M N, December 8, 2021
https://cryptobullsclub.com/top-ethereum-alternatives/

41 Understanding Smart Contracts – Panel at DC Blockchain 2016
https://www.youtube.com/watch?v=ZuHZOryZ_f0/

Chapter 6 – Enterprise, Hyperledger Fabric and Modularity

42 Hyperledger Project, February 2016 https://www.hyperledger.org/announcements/2016/02/09/linux-foundations-hyperledger-project-announces-30-founding-members-and-code-proposals-to-advance-blockchain-technology

43 Hyperledger Fabric, Jake Frankenfield, September 23, 2021
https://www.investopedia.com/terms/h/hyperledger-fabric.asp/

44 Does Hyperledger Fabric Perform at Scale? Christopher Ferris, April 2018
https://www.ibm.com/blogs/blockchain/2019/04/

does-hyperledger-fabric-perform-at-scale/

45   FastFabric: Scaling Hyperledger Fabric to 20,000 Transactions per Second, Christian Gorenflo, Stephen Lee, Lukasz Golab, S. Keshav, March 2019 https://arxiv.org/abs/1901.00910/

Chapter 7 - Artificial Intelligence, IoT and Quantum

46   "Combining Blockchain and Artificial Intelligence for a Better Future." Blog. modex.tech, Modex Team, August 13, 2018 http://blog.modex.tech/combining-blockchain-and- artificial-intelligence-for-a-better-future-421e97141e60/

47   Blockchain Serves as Tool for Human, Product and IoT Device Identity Validation, Chain of Things Limited, Hans Lombardo, January 11, 2017 https://inform.tmforum.org/features-and-analysis/2016/11/blockchain-serves-tool-human-product-iot-device-identity-validation/

48   How Blockchain and Smart Contracts Can Impact IoT, Smartz Platform Blog: Medium, Smartz, August 21, 2018 http://medium.com/smartz-blog/how-blockchain-and- smart-contracts-can-impact-iot-f9e77ebe02ab/

49   The Future of Energy Is Local, Brooklyn Microgrid, 2018 http://brooklynmicrogrid.com/

50   The Future of Energy: Blockchain, Transactive Grids, Microgrids, Energy Trading: LO3 Stock, Tokens and Information. LO3 Energy, 2018. http://www.lo3energy.com/

Chapter 8 - Cybersecurity, Zero Knowledge Proofs, Digital Identity

51   2022 SonicWall Cyber Threat Report https://www.sonicwall.com/2022-cyber-threat-report/

52   How Blockchain Could Revolutionize Cybersecurity, Robert Napoli, March 2022 https://www.forbes.com/sites/forbestechcouncil/2022/03/04/ how-blockchain-could-revolutionize-cybersecurity/

53   Decentralized naming and certificate authority, Handshake https://handshake.org/

54   An Overview of Decentralized Cloud Storage Services, Cryptopedia Staff,

December 21, 2021
https://www.gemini.com/cryptopedia/crypto-cloud-storage-decentralized-cloud-storage-providers

55   Applied Kid Cryptography or How to Convince Your Children You Are Not Cheating, Moni Naor, Yael Naor, Omer Reingold, March 1999 https://www.wisdom.weizmann.ac.il/~naor/PAPERS/waldo.pdf

56   How Zero-Knowledge Proof Enables Trustless Transactions and Increases Your Privacy, Dr Mark van Rijmenam, December 2017 https://www.thedigitalspeaker.com/zero-knowledge-proof-enables-trustless-transactions-increases-privacy/

57   A Gentle Introduction to Verifiable Credentials, Daniel Hardman, October 2019 https://www.evernym.com/blog/gentle-introduction-verifiable-credentials/

58   Verifiable Credentials, Wikipedia https://en.wikipedia.org/wiki/Verifiable_credentials

59   Why Software is No Longer Being Written from Scratch, Richard Harris, November 2016, https://appdevelopermagazine.com/why-software-is-no-longer-being-written-from-scratch/

60   The Linux Foundation Releases The State of Software Bill of Materials (SBOM) and Cybersecurity Readiness Research, February 2022 https://www.linuxfoundation.org/press-release/the-linux-foundation-releases-the-state-of-software-bill-of-materials-sbom-and-cybersecurity-readiness-research/

61   Securing the Nation's Software Supply Chain, September 17, 2021 https://dragonchain.com/blog/secure-the-nations-software-supply-chain

62   Why the World Needs a Software Bill Of Materials Now, Dr. Sybe Izaak Rispens, Mar 14, 2021 https://drrispens.medium.com/why-the-world-needs-a-software-bill-of-materials-now-5a565df65dff

63   A Software Bill of Materials Is Critical for Comprehensive Risk Management, Dr. Georgianna Shea, September 2021 https://www.fdd.org/wp-content/uploads/2021/09/fdd-memo-a-software-bill-of-materials-is-critical-for-comprehensive-risk-management.pdf

64   The Future Use Cases of Blockchain for Cybersecurity, Julien Legrand,

September 2020
https://www.cm-alliance.com/cybersecurity-blog/the-
future-use-cases-of-blockchain-for-cybersecurity

Chapter 9 - A Road to Web3

65   Solidarity and Betrayal: The Rise and Fall of the Players' League, Emma
      Baccellieri, Jan 2022
      https://www.si.com/mlb/2022/01/05/players-league-daily-cover

66   Web3, Wikipedia.org
      https://en.wikipedia.org/wiki/Web3

67   What is Web3, Kevin Roose
      https://www.nytimes.com/interactive/2022/03/18/
      technology/Web3-definition-internet.html

68   Hybrid Smart Contracts Explained, Chainlink, May 2021
      https://blog.chain.link/hybrid-smart-contracts-explained/

69   State (computer science), Wikipedia
      https://en.wikipedia.org/wiki/State_(computer_science)

70   Understanding Web 3 — A User Controlled Internet, Emre Tekisalp, August
      2018
      https://blog.coinbase.com/understanding-web-3-
      a-user-controlled-internet-a39c21cf83f3

71   Festivus! The Website
      https://festivusweb.com/festivus-airing-of-grievances.php

72   Better by the Dozen: Bringing Back the Twelve-Person Civil Jury, Gensler,
      Higginbotham, Rosenthal, Summer 2020
      https://judicature.duke.edu/articles/better-by-the-dozen-
      bringing-back-the-twelve-person-civil-jury/

Addendum 2 - I'm Satoshi?

73   Who is the Mysterious Bitcoin Creator Satoshi Nakamoto?
      https://cointelegraph.com/bitcoin-for-beginners/who-
      is-satoshi-nakamoto-the-creator-of-bitcoin

74   P2P Foundation
      http://p2pfoundation.ning.com/forum/topics/bitcoin-open-so
      urce?commentId=2003008%3AComment%3A52186

75   Unmasking Satoshi Nakamoto, Who is He? Watcher.Guru, September 8, 2021
https://watcher.guru/news/unmasking-satoshi-nakamoto-who-is-he

76   The Many Facts Pointing to Hal Finney Being Satoshi Nakamoto, Graham Smith
Nov 20, 2019
https://news.bitcoin.com/the-many-facts-pointing-to-hal-finney-being-satoshi/

77   Anonymous Remailer, Wikipedia.com
https://en.wikipedia.org/wiki/Anonymous_remailer/

78   Is Crypto Expert Len Sassaman the Creator of Bitcoin, Satoshi Nakamoto?
March 2021
https://www.cnbctv18.com/market/is-crypto-expert-len-sassaman-the-creator-of-bitcoin-satoshi-nakamoto-8520761.htm